Green Energy and Technology

Ayhan Demirbas · M. Fatih Demirbas

Algae Energy

Algae as a New Source of Biodiesel

 Springer

Ayhan Demirbas
Professor of Energy Technology
Sirnak University
Sirnak
Turkey

M. Fatih Demirbas
Sila Science and Energy Unlimited Company
University Mah.
24 Akif Saruhan Cad., Mekan Sok.
61080 Trabzon
Turkey

ISSN 1865-3529
ISBN 978-1-84996-049-6   e-ISBN 978-1-84996-050-2
DOI 10.1007/978-1-84996-050-2
Springer London Dordrecht Heidelberg New York

British Library Cataloguing in Publication Data
A catalogue record for this book is available from the British Library

Library of Congress Control Number: 2010929649

© Springer-Verlag London Limited 2010
Apart from any fair dealing for the purposes of research or private study, or criticism or review, as permitted under the Copyright, Designs and Patents Act 1988, this publication may only be reproduced, stored or transmitted, in any form or by any means, with the prior permission in writing of the publishers, or in the case of reprographic reproduction in accordance with the terms of licenses issued by the Copyright Licensing Agency. Enquiries concerning reproduction outside those terms should be sent to the publishers.
The use of registered names, trademarks, etc., in this publication does not imply, even in the absence of a specific statement, that such names are exempt from the relevant laws and regulations and therefore free for general use.
The publisher and the authors make no representation, express or implied, with regard to the accuracy of the information contained in this book and cannot accept any legal responsibility or liability for any errors or omissions that may be made.

Cover design: WMXDesign, Heidelberg, Germany

Printed on acid-free paper

Springer is part of Springer Science+Business Media (www.springer.com)

Preface

This book examines the production of algae culture and usage of algal biomass conversion products. In this book, the modern biomass-based transportation fuels biodiesel, bio-oil, biomethane, biohydrogen, and high-value-added products from algae are briefly reviewed. The most significant distinguishing characteristic of algal oil is its yield and, hence, its biodiesel yield. According to some estimates, the yield (per acre) of oil from algae is over 200 times the yield from the best-performing plant/vegetable oils. The lipid and fatty acid contents of microalgae vary in accordance with culture conditions. The availability of algae and the advantages of algal oil for biodiesel production have been investigated.

Billions of years ago the Earth's atmosphere was filled with CO_2. Thus there was no life on Earth. Life on Earth started with *Cyanobacteria* and algae. These humble photosynthetic organisms sucked out the atmospheric CO_2 and started releasing oxygen. As a result, the levels of CO_2 started decreasing to such an extent that life evolved on Earth. Once again these smallest of organisms are poised to save us from the threat of global warming.

In the context of climatic changes and soaring prices for a barrel of petroleum, biofuels are now being presented as a renewable energy alternative. Presently, research is being done on microscopic algae, or microalgae, which are particularly rich in oils and whose yield per hectare is considerably higher than that of sunflower or rapeseed. Algae will become the most important biofuel source in the near future. Microalgae appear to be the only source of renewable biodiesel that is capable of meeting the global demand for transport fuels. Microalgae are theoretically a very promising source of biodiesel.

Algae are the fastest-growing plants in the world. Industrial reactors for algal culture are open ponds, photobioreactors, and closed systems. Algae are very important as a biomass source and will some day be competitive as a source for biofuel. Different species of algae may be better suited for different types of fuel. Algae can be grown almost anywhere, even on sewage or salt water, and does not require fertile land or food crops, and processing requires less energy than the algae provides. Algae can be a replacement for oil-based fuels, one that is more effective and has no disadvantages. Algae are among the fastest-growing plants in the world, and about 50% of their weight is oil. This lipid oil can be used to make biodiesel for cars, trucks, and airplanes. Microalgae have much faster growth rates than terrestrial crops. the per unit area yield of oil from algae is estimated to be between 20,000 and 80,000 L per acre per year; this is 7 to 31 times greater than the next best crop, palm oil. Most current research on oil extraction is focused on microalgae to produce biodiesel from algal oil. Algal oil is processed into biodiesel as easily as oil derived from land-based crops. Algae biomass can play an important role in solving the problem of food or biofuels in the near future.

Microalgae contain oils, or lipids, that can be converted into biodiesel. The idea of using microalgae to produce fuel is not new, but it has received renewed attention recently in the search for sustainable energy. Biodiesel is typically produced from plant oils, but there are widely voiced concerns about the sustainability of this practice. Biodiesel produced from microalgae is being investigated as an alternative to using conventional crops such as rapeseed; microalgae typically produce more oil, consume less space, and could be grown on land unsuitable for agriculture.

Using microalgae as a source of biofuels could mean that enormous cultures of algae are grown for commercial production, which would require large quantities of fertilizers. While microalgae are estimated to be capable of producing 10 to 20 times more biodiesel than rapeseed, they need 55 to 111 times more nitrogen fertilizer – 8 to 16 tons/ha/year.

This book on algae energy attempts to address the needs of energy researchers, chemical engineers, chemical engineering students, energy resource specialists, engineers, agriculturists, crop cultivators, and others interested in practical tools for pursuing their interests in relation to bioenergy. Each chapter in the book starts with basic explanations suitable for general readers and ends with in-depth scientific details suitable for expert readers. General readers include people interested in learning about solutions to current fuel and environmental crises. Expert readers include chemists, chemical engineers, fuel engineers, agricultural engineers, farming specialists, biologists, fuel processors, policymakers, environmentalists, environmental engineers, automobile engineers, college students, research faculty, etc. The book may even be adopted as a textbook for college courses that deal with renewable energy or sustainability.

Trabzon, TURKEY (September 2009)

Ayhan Demirbas
Muhammet Demirbas

Contents

1 Introduction 1
 1.1 Introduction 1
 1.2 Energy Demand and Availability 2
  1.2.1 Fossil Energy Sources 3
  1.2.2 Renewable Energy Sources 10
 1.3 Present Energy Situation 12
  1.3.1 Energy Production and Future Energy Scenarios 12
  1.3.2 Future Energy Scenarios 13
 References 26

2 Green Energy Facilities 29
 2.1 Introduction 29
 2.2 Biomass Feedstocks 30
 2.3 Green Energy 31
 2.4 Importance of Biomass 32
 2.5 Production of Fuels and Chemicals from Biomass 36
 References 46

3 Biofuels 49
 3.1 Introduction 49
 3.2 Importance of Biofuels 51
 3.3 Bioethanol 54
 3.4 Biodiesel 56
 3.5 Bio-oil 59
 3.6 Biogas 61
 3.7 Fischer–Tropsh Liquids from Biorenewable Feedstocks 64
 3.8 Biohydrogen 69
 3.9 Other Liquid Biofuels 70
  3.9.1 Glycerol-based Fuel Oxygenates for Biodiesel
    and Diesel Fuel Blends 70
  3.9.2 P-series Fuels 71

vii

Contents

3.9.3 Dimethyl Ether 71
3.9.4 Other Bio-oxygenated Liquid Fuels 72
References 72

4 Algae Technology 75
4.1 Introduction 75
    4.1.1 Definition of Algal Terms 76
4.2 Production Systems.. 77
    4.2.1 Harvesting Microalgae............................... 79
    4.2.2 Photobioreactors.................................... 80
    4.2.3 Open-pond Systems 83
    4.2.4 Closed and Hybrid Systems 87
4.3 Production Costs 89
References 93

5 Energy from Algae 97
5.1 Introduction 97
    5.1.1 Historical Perspective 99
5.2 Biofuels 99
    5.2.1 Biodiesel 102
    5.2.2 Bioalcohol 104
    5.2.3 Costs, Prices, and Economic Impacts of Biofuels.......... 106
    5.2.4 Environmental Impacts of Biofuels 110
    5.2.5 Combustion Efficiencies of Biofuels 113
    5.2.6 Bio-oil 117
    5.2.7 Biomethane 119
    5.2.8 Production of Bio-oil and Hydrogen by Pyrolysis 120
    5.2.9 Anaerobic Biohydrogen Production 125
5.3 Liquefaction of Algal Cells 127
    5.3.1 Liquefaction of Algal Cells by Hexane Extraction 129
5.4 High-value-added Products from Algae 131
    5.4.1 Small Molecules 131
    5.4.2 Polymers 132
    5.4.3 High-value Oils..................................... 132
References 133

6 Biodiesel from Algae 139
6.1 Introduction 139
6.2 Biodiesel from Algal Oil 140
    6.2.1 Production of Biodiesel from Algal Oils 144
6.3 Potential of Microalgal Biodiesel 147
6.4 Acceptability of Microalgal Biodiesel 150
6.5 Economics of Biodiesel Production 151
6.6 Improving Economics of Microalgal Biodiesel 153
6.7 Advantages and Disadvantages of Biodiesel from Algal Oil 153
References 154

Contents

ix

7 Biorefineries 159
   7.1 Introduction 159
   7.2 Definitions of Biorefinery 162
      7.2.1 Main Technical and Nontechnical Gaps
         and Barriers to Biorefineries 164
   7.3 Historical Perspective 164
   7.4 Petroleum Refinery and Biorefinery 168
   7.5 Refining of Upgraded Products 174
   7.6 Opportunities for Refining Pyrolysis Products 178
   References 180

8 Future Developments 183
   8.1 Introduction 183
      8.1.1 World Theoretical Limit of Biomass Supply 184
      8.1.2 High-yield Energy Crops 185
      8.1.3 Food Versus Fuel Delineation 186
      8.1.4 Thermodynamic Efficiency (Exergy Analysis) 187
      8.1.5 Biofuel Upgradation 187
      8.1.6 Carbon Credits 188
   8.2 Social and Political Issues 189
      8.2.1 The Promise of Algae: Energy Security 190
   8.3 Environmental Impacts of Biomass Production 191
   References 193

Index 195

Chapter 1
Introduction

1.1 Introduction

Energy is defined as the ability to do work. However, there is no concept of force that includes all definitions of force. For example, the forces that form light energy, heat (thermal) energy, mechanical energy, electrical energy, magnetic energy, etc. are different. There are many forms of energy, but they all fall into one of two categories: kinetic and potential. Electrical, radiant, thermal, motion, and sound energies are kinetic; chemical, stored mechanical, nuclear, and gravitational energies are types of potential energy. It can exist in the form of motion. This is known as kinetic energy. The motion can relate to different things. If the motion is of a large object, the kinetic energy is said to be mechanical. If moving objects are electrically charged, they are said to form an electric current. If moving objects are individual molecules, there are two possibilities. If their motion is organized into waves, then their kinetic energy is associated with sound. If their motion is completely disorganized, then their kinetic energy is associated with what we call heat or thermal energy. Another form of kinetic energy is light (and other forms of electromagnetic radiation, like radio waves and microwaves). Other forms of energy do not have the form of motion, but they can cause an increase in motion at a later time. Water at the top of a dam can spill over the dam. A battery can produce an electric current when it is connected to a circuit. Fuels can be burned to produce heat. All of these are examples of potential energy.

The world is presently confronted with twin crises of fossil fuel depletion and environmental degradation. To overcome these problems, renewable energy has recently been receiving increased attention due to its environmental benefits and the fact that it is derived from renewable sources such as virgin or cooked vegetable oils (both edible and nonedible). The world's excessive demand for energy, the oil crisis, and the continuous increase in oil prices have led countries to investigate new and renewable fuel alternatives. Hence, energy sources such as solar, wind, geothermal, hydro, nuclear, hydrogen, and biomass have been considered.

A. Demirbas, M. Fatih Demirbas, *Algae Energy*
DOI 10.1007/978-1-84996-050-2, © Springer 2010

1.2 Energy Demand and Availability

Energy plays a vital role in our everyday lives. A country's standard of living is considered to be proportional to the energy consumption by the people of that country. Energy is one of the vital inputs to the socioeconomic development of any country. The abundance of energy around us can be stored, converted, and amplified for our use in a variety of ways. Energy production has always been a concern for researchers as well as policymakers.

Global energy sources are classified into two groups, fossil and renewable. Primary energy sources can be divided into nonrenewables and renewables. Nonrenewable energy sources include coal, petroleum, gas, gas hydrate, and fissile material, while renewable energy sources include biomass, hydro, geothermal, solar, and wind energy. The main fissile energy sources are uranium and thorium.

An energy source can also be classified according to its depletion rate. While biomass energy can be depleted, solar and wind energy are nondepletable. In reality the energy availability from nonrenewable sources is limited, and beyond that, the exploration, processing, and use of energy involve considerable impacts on the environment. Fossil fuels have been the prime sources of energy for the purpose of transportation, power generation, and agriculture, as well as in commercial, residential, and industrial activity for more than a century. The world's energy requirements are currently satisfied by fossil fuels, which serve as the primary energy source.

Because of the increase in petroleum prices, especially after the oil crisis in 1973 and then the Gulf war in 1991, in addition to the geographically reduced availability of petroleum and more stringent governmental regulations on exhaust emissions, researchers have studied alternative fuels and alternative solutions.

Interestingly, renewable energy resources are more evenly distributed than fossil or nuclear resources. Today's energy system is unsustainable because of equity issues as well as environmental, economic, and geopolitical concerns that will have implications far into the future. Hence, sustainable renewable energy sources such as biomass, hydro, wind, solar (both thermal and photovoltaic), geothermal, and marine energy sources will play an important role in the world's future energy supply.

Developing renewable sources of energy has become necessary due to the limited supply of fossil fuels. Global environmental concerns and decreasing resources of crude oil have prompted demand for alternative fuels. Global climate change is also the major environmental issue of our time. Global warming, the Kyoto Protocol, the emission of greenhouse gases, and the depletion of fossil fuels are the topics of environmental concern worldwide. Due to rapidly increasing energy requirements along with technological development around the world, research and development activities have perforce focused on new and renewable energy.

The major sources of alternative energy are biorenewables, hydro, solar, wind, geothermal, and other forms of energy, each of them having their own advantages and disadvantages, including political, economic, and practical issues. Renewable energy is a promising alternative solution because it is clean and environmentally safe. Sources of renewable energy also produce lower or negligible levels of green-

1.2 Energy Demand and Availability

house gases and other pollutants as compared with the fossil energy sources they replace.

1.2.1 Fossil Energy Sources

The term fossil refers to an earlier geologic age. Fossil fuels were formed a long time ago and are not renewable. Fossil energy sources are petroleum (crude oil), coal, bitumens, natural gas, oil shales, and tar sands. During the last 200 years, developed countries have shifted their energy consumption toward fossil fuels. About 98% of carbon emissions result from fossil fuel combustion. Reducing the use of fossil fuels would considerably reduce the amount of carbon dioxide and other pollutants produced. In fact, today over 80% of the energy we use comes from three fossil fuels: petroleum, coal, and natural gas. Unfortunately, oil is in danger of becoming scarce.

Another problem with petroleum fuels is their uneven distribution in the world; for example, the Middle East has 63% of the global reserves and is the dominant supplier of petroleum. This energy system is unsustainable because of equity issues as well as environmental, economic, and geopolitical concerns that have far-reaching implications.

The current global energy mix consists of oil (36%), natural gas (24%), coal (28%), nuclear (6%), and renewable energy such as hydro, wind, and solar (about 7%). Once the energy picture has been established we will explore the effect the projected changes in energy supply may have on the world population. Petroleum is the largest single source of energy consumed by the world's population, exceeding coal, natural gas, nuclear, hydro, and renewables. While fossil fuels are still being created today by underground heat and pressure, they are being consumed more rapidly than they are being created. Hence, fossil fuels are considered nonrenewable; that is, they are not replaced as fast as they are consumed. And due to oil's aforementioned looming scarcity, the future trend is toward using alternative energy sources. Fortunately, the technological advances are making the transition possible (Kirtay 2009).

The word petroleum comes from the Greek word *petra*, or rock, and Latin word *oleum*, oil. Oil is a thick, dark brown or greenish liquid found in reservoirs in sedimentary rock. Tiny pores in the rock allowed the petroleum to seep in. These "reservoir rocks" hold the oil like a sponge, confined by other, nonporous layers that form a trap. Petroleum is used to describe a broad range of hydrocarbons that are found as gases, liquids, or solids beneath the surface of the Earth. The two most common forms are natural gas and crude oil. Petroleum consists of a complex mixture of various hydrocarbons, largely of alkane and aromatic compounds. The color ranges from pale yellow through red and brown to black or greenish, while by reflected light it is, in the majority of cases, of a green hue. Petroleum is a fossil fuel because it was formed from the remains of tiny sea plants and animals that died millions of years ago and sank to the bottom of the oceans.

4   1 Introduction

Table 1.1 shows crude oil production data for various regions (IEA 2007). The Middle East produces 32% of the world's oil, and, more importantly, it has 64% of the total proven oil reserves in the world. Oil fields follow a size distribution consisting of a very few large fields and many smaller ones. This distribution is illustrated by the fact that 60% of the world's oil supply is extracted from only 1% of the world's active oil fields. As one of these very large fields plays out it can require the development of hundreds of small fields to replace its production.

Some definitions will be useful. "Petroleum" and "oil" are used interchangeably to include crude oil, shale oil, oil sands, and natural gas liquids (NGLs). The word petroleum generally refers to crude oil or the refined products obtained from the processing of crude oil (gasoline, diesel fuel, heating oil, etc.). Crude oil (raw petroleum) is separated into fractions by fractional distillation. The fractions at the top are lower than those at the bottom. The heavy bottom fractions are often cracked into lighter, more useful products. All of the fractions are processed further in other refining units. The main crude oil fractions are shown in Table 1.2.

Crude oil is separated by boiling points into six main grades of hydrocarbons: refinery gas (used for refinery fuel), gasoline (naphthas), kerosene, light oils (diesel oil or diesel fuel) and heavy gas oils (fuel oil), and long residue. This initial sep-

Table 1.1 1973 and 2006 regional shares of crude oil production (%)

Region	1973	2006
Middle East	37.0	31.1
OECD	23.6	23.2
Former USSR	15.0	15.2
Africa	10.0	12.1
Latin America	8.6	9.0
Asia excluding China	3.2	4.5
China	1.9	4.7
Non-OECD Europe	0.7	0.2
Total (Millions of tons)	2,867	3,936

Table 1.2 Main crude oil fractions

Component	Boiling range, K	Number of carbon atoms
Natural gas	< 273	C_1 to C_4
Liquefied petroleum gas	231–273	C_3 to C_4
Petroleum ether	293–333	C_5 to C_6
Ligroin (light naphtha)	333–373	C_6 to C_7
Gasoline	313–478	C_5 to C_{12}, and cycloalkanes
Jet fuel	378–538	C_8 to C_{14}, and aromatics
Kerosene	423–588	C_{10} to C_{16}, and aromatics
No. 2 diesel fuel	448–638	C_{10} to C_{22}, and aromatics
Fuel oils	> 548	C_{12} to C_{70}, and aromatics
Lubricating oils	> 673	$> C_{20}$
Asphalt or petroleum coke	Nonvolatile residue	Polycyclic structures

1.2 Energy Demand and Availability

aration is done by distillation. The first step in the refining of crude oil, whether in a simple or a complex refinery, is the separation of the crude oil into fractions (fractionation or distillation). These fractions are mixtures containing hydrocarbon compounds whose boiling points lie within a specified range.

A diesel engine burns fuel oil rather than gasoline and differs from the gasoline engine in that it uses compressed air in the cylinder rather than a spark to ignite the fuel. Diesel or diesel fuel in general is any fuel used in diesel engines. Diesel engines are used mainly in heavy vehicles. The main advantage of the diesel engine is that the level of efficiency is greater than in the Otto cycle engine. This means that a greater part of the energy content of the fuel is used. The efficiency of a diesel engine is at best 45%, compared to 30% for the Otto engine.

Diesel fuel is produced by distilling raw oil extracted from bedrock. Diesel is a fossil fuel. Diesel fuel consists of hydrocarbons with between 9 and 27 carbon atoms in a chain as well as a smaller amount of sulfur, nitrogen, oxygen, and metal compounds. It is a general property of hydrocarbons that the auto-ignition temperature is higher for more volatile hydrocarbons. The hydrocarbons present in diesel fuels include alkanes, naphthenes, olefins, and aromatics. In addition, other substances are added to improve the characteristics of diesel fuel. Its boiling point is between 445 and 640 K. A good diesel fuel is characterized by low sulfur and aromatic content, good ignition quality, the right cold weather properties, and a low content of pollutants, as well as the right density, viscosity, and boiling point.

Diesel fuel comes in several different grades, depending upon its intended use. Like gasoline, diesel fuel is not a single substance but a mixture of various petroleum-derived components, including paraffins, isoparaffins, napthenes, olefins, and aromatic hydrocarbons, each with their own physical and chemical properties.

Unlike spark-ignition engines, the power and economy of diesel engines are comparatively insensitive to fuel volatility. There is some indirect impact in that less volatile fuels have higher heating values (HHVs). Conversely, fuels with higher front-end volatility tend to improve starting and warm-up performance and reduce smoke. Ideal fuel volatility requirements will vary based on engine size and design, speed and load conditions, and atmospheric conditions. As an example, more volatile fuels may provide better performance for fluctuating loads and speeds such as those experienced by trucks and buses.

The viscosity of diesel fuel is an important property that impacts the performance of fuel injection systems. Some injection pumps can experience excessive wear and power loss due to injector or pump leakage if viscosity is too low. If fuel viscosity is too high, it may cause too much pump resistance and filter damage and adversely affect fuel spray patterns. High fuel viscosity can cause an injector spray pattern with poor fuel dispersion.

Gasoline is a petroleum-derived liquid mixture, primarily used as fuel in internal combustion engines, specifically in spark-ignition engines. In the Otto cycle engine a mixture of gasoline and air is compressed and then ignited by a spark plug.

The important characteristics of gasoline are density, vapor pressure, distillation range, octane number, and chemical composition. To be attractive, a motor gasoline must have (a) the desired volatility, (b) antiknock resistance (related to octane rating), (c) good fuel economy, (d) minimal deposition on engine component surfaces, and (e) complete combustion and low pollutant emissions.

The density of gasoline is 0.71 to 0.77 kg/L. Gasoline is more volatile than diesel oil, Jet-A, or kerosene, not only because of its base constituents but because of the additives that are put into it. The final control of volatility is often achieved by blending with butane. The desired volatility depends on the ambient temperature: in hotter climates, gasoline components of higher molecular weight, and thus lower volatility, are used. In cold climates, too little volatility results in cars failing to start. In hot climates, excessive volatility results in what is known as "vapor lock," where combustion fails to occur because the liquid fuel has changed into a gaseous fuel in the fuel lines, rendering the fuel pump ineffective and starving the engine of fuel.

An important characteristic of gasoline is its octane number or octane rating, which is a measure of how resistant gasoline is to the abnormal combustion phenomenon known as predetonation (also known as knocking, pinging, spark knock, and other names). Octane number is measured relative to a mixture of 2,2,4-trimethylpentane and n-heptane. Octane number is a measure of the gasoline quality for the prevention of early ignition, which leads to cylinder knocks. Higher octane numbers are preferred in internal combustion engines. For gasoline production, aromatics, naphthenes, and isoalkanes are highly desirable, whereas olefins and n-paraffins are less desirable.

The typical composition of gasoline hydrocarbons (% volume) is as follows: 4 to 8% alkanes, 2 to 5% alkenes, 25 to 40% isoalkanes, 3 to 7% cycloalkanes, to 4% cycloalkenes, and 20 to 50% total aromatics (0.5 to 2.5% benzene). Ad-

Table 1.3 Physical and chemical properties of gasoline

Property	Information
Color	Colorless to pale brown or pink
Average molecular weight	108
Density, kg/L	0.7–0.8
Flash point, K	227.2
Explosive limits in air	1.3–6.0%
Flammability limits	1.4–7.4%
Autoignition, K	553–759
Boiling point, K	
Initially	312
After 10% distillate	333
After 50% distillate	383
After 90% distillate	443
Final boiling point	477
Solubility	
Water at 293 K	Insoluble
Absolute ethanol	Soluble
Diethyl ether	Soluble
Chloroform	Soluble
Benzene	Soluble

1.2 Energy Demand and Availability

ditives and blending agents are added to the hydrocarbon mixture to improve the performance and stability of gasoline. These compounds include antiknock agents, antioxidants, metal deactivators, lead scavengers, antirust agents, anti-icing agents, upper-cylinder lubricants, detergents, and dyes. The physical and chemical properties of gasoline are given in Table 1.3. Table 1.4 shows the major components of gasoline.

Worldwide coal production is roughly equal to gas production and only second to that of oil. Coal is produced in deep mines (hard coal) and in surface mines (lignite). Coal has played a key role as a primary source of organic chemicals as well as a primary energy source. Coal may become more important both as an energy source and as the source of carbon-based materials, especially aromatic chemicals, in the 21st century (Schobert and Song 2002).

Table 1.4 Major components of gasoline

Component	Composition, % by weight
n-alkanes	
C_5	3.0
C_6	11.6
C_7	1.2
C_9	0.7
C_{10}–C_{13}	0.8
Total n-alkanes	17.3
Branched alkanes	
C_4	2.2
C_5	15.1
C_6	8.0
C_7	1.9
C_8	1.8
C_9	2.1
C_{10}–C_{13}	1.0
Total branched alkanes	32.0
Cycloalkanes	
C_6	3.0
C_7	1.4
C_8	0.6
Total cycloalkanes	5.0
Olefins	
C_6	1.8
Total olefins	1.8
Aromatics	
Benzene	3.2
Toluene	4.8
Xylenes	6.6
Ethylbenzene	1.4
C_3-benzenes	4.2
C_4-benzenes	7.6
Others	2.7
Total aromatics	30.5

The first known and oldest fossil fuel is coal. Coal has played a key role as a primary energy source as well as a primary source of organic chemicals. It is a complex, heterogeneous combustible material made up of portions that are either useful (carbon and hydrogen) or useless (diluents such as moisture, ash, and oxygen or contaminants such as sulfur and heavy metals). Coal can be defined as a sedimentary rock that burns. It was formed by the decomposition of plant matter and is a complex substance that can be found in many forms. Coal is divided into four classes: anthracite, bituminous, subbituminous, and lignite. Elemental analysis gives empirical formulas such as $C_{137}H_{97}O_9NS$ for bituminous coal and $C_{240}H_{90}O_4NS$ for high-grade anthracite.

Coal accounts for 26% of the world's primary energy consumption and 37% of the energy consumed worldwide for electricity generation. For coal to remain competitive with other sources of energy in the industrialized countries of the world, continuing technological improvements in all aspects of coal extraction are necessary. Nearly all the different forms of coal are used in one way or another. For instance, peat has been used for burning in furnaces, lignite is used in power stations and home (residential) stoves, whereas bituminous coal is used extensively for the generation of electricity.

Coal is formed from plant remains that have been compacted, hardened, chemically altered, and metamorphosed underground by heat and pressure over millions of years. When plants die in a low-oxygen swamp environment, instead of decaying by bacteria and oxidation, their organic matter is preserved. Over time, heat and pressure remove the water and transform the matter into coal. The first step in coal formation yields peat, compressed plant matter that still contains leaves and twigs. The second step is the formation of brown coal or lignite. Lignite has already lost most of the original moisture, oxygen, and nitrogen. It is widely used as a heating fuel but is of little chemical interest. The third stage, bituminous coal, is also widely utilized as a fuel for heating. Bituminous is the most abundant form of coal and is the source of coke for smelting, coal tar, and many forms of chemically modified fuels. Table 1.5 shows the world's recoverable coal reserves (IEA 2007).

The role of natural gas (NG) in the world's energy supply is growing rapidly. NG is the fastest growing primary energy source in the world. The reserves and resources of conventional NG are comparable in size to those of conventional oil, but global gas consumption is still considerably lower than that of oil. Proven gas reserves are not evenly distributed around the globe: 41% are in the Middle East and 27% in Russia. A peak in conventional gas production may occur between 2020 and 2050. NG accounts today for 25% of world primary energy production (Jean-Baptiste and Ducroux 2003). Because it is cleaner fuel than oil or coal and not as controversial as nuclear power, gas is expected to be the fuel of choice for many countries in the future. Increasing demand for NG is expected in all sectors of the world, as resource availability, rate of depletion, and environmental considerations all favor its use. World NG reserves by country are given in Table 1.6.

1.2 Energy Demand and Availability 9

Table 1.5 World's recoverable coal reserves

Country	Bituminous including anthracite	Subbituminous	Lignite
United States	115,891	101,021	33,082
China	62,200	33,700	18,600
India	82,396	–	2,000
South Africa	49,520	–	–
Kazakhstan	31,100	–	3,000
Brazil	–	11,929	–
Colombia	6,267	381	–
Canada	3,471	871	2,236
Indonesia	790	1,430	3,150
Botswana	4,300	–	–
Uzbekistan	1,000	–	3,000
Turkey	278	761	2,650
Pakistan	–	2,265	–
Thailand	–	–	1,268
Chile	31	1,150	–
Mexico	860	300	51
Peru	960	–	100
Kyrgyzstan	–	–	812
Japan	773	–	–
Korea (Dem. People's Rep.)	300	300	–
Zimbabwe	502	–	–
Venezuela	479	–	–
Philippines	–	232	100
Mozambique	212	–	–
Swaziland	208	–	–
Tanzania	200	–	–
Others	449	379	27

Table 1.6 World natural gas reserves by country

Country	Percent of world total
Russian Federation	33.0
Iran	15.8
Qatr	5.8
United Arab Emirates	4.1
Saudi Arabia	4.0
United States	3.3
Venezuela	2.8
Algeria	2.5
Nigeria	2.4
Iraq	2.1
Turkmenistan	2.0
Top 20 countries	89.0
Rest of world	11.0

1.2.2 Renewable Energy Sources

Renewable energy sources that use indigenous resources have the potential to provide energy services with zero or almost zero emissions of both air pollutants and greenhouse gases. Currently, renewable energy sources supply 14% of the total world energy demand. Large-scale hydropower supplies 20% of global electricity. Renewable resources are more evenly distributed than fossil and nuclear resources. Renewable energy scenarios depend on environmental protection, which is an essential characteristic of sustainable developments.

For biomass resources, several potential sources may be used. Biomass resources include agricultural and forest residues, algae and grasses, animal manure, organic wastes, and biomaterials. The supply is dominated by traditional biomass used for cooking and heating, especially in rural areas of developing countries. Worldwide biomass ranks fourth as an energy resource, providing approximately 14% of the world's energy needs (Hall *et al.* 1992).

Biomass now represents only 3% of primary energy consumption in industrialized countries. However, much of the rural population in developing countries, which represents about 50% of the world's population, is reliant on biomass, mainly in the form of wood, for fuel (Ramage and Scurlock 1996). In Europe, North America, and the Middle East, the share of biomass averages 2 to 3% of total final energy consumption, whereas in Africa, Asia, and Latin America, which together account for three-quarters of the world's population, biomass provides a substantial share of energy needs: a third on average, but as much as 80 to 90% in some of the poorest countries of Africa and Asia (e.g., Angola, Ethiopia, Mozambique, Tanzania, Democratic Republic of Congo, Nepal, and Myanmar). Large-scale hydropower provides about one-quarter of the world's total electricity supply, virtually all of Norway's electricity, and more than 40% of the electricity used in developing countries. The technically usable world potential of large-scale hydro is estimated to be over 2,200 GW.

There are two small-scale hydropower systems: micro hydropower systems (MHP), with capacities below 100 kW, and small hydropower systems (SHP), with capacity between 101 kW and 1MW. Large-scale hydropower supplies 20% of global electricity. According to the United Nations Development Programme, in developing countries, considerable potential still exists, but large hydropower projects may face financial, environmental, and social constraints (UNDP 2000).

Geothermal energy for electricity generation has been produced commercially since 1913 and for four decades on the scale of hundreds of megawatts both for electricity generation and direct use. Utilization of geothermal sources of energy has increased rapidly during the last three decades. In 2000, geothermal resources were identified in over 80 countries, and there are quantified records of geothermal utilization in 58 countries in the world. Table 1.7 shows the status of geothermal energy (Fridleifsson 2001).

Geothermal energy is clean, cheap, and renewable and can be utilized in various forms such as space heating and domestic hot water supply, CO_2 and dry-ice production processes, heat pumps, greenhouse heating, swimming and balneology

1.2 Energy Demand and Availability

11

Table 1.7 World's top countries using geothermal energy directly

Country	Installed MWt	Production (GWh/a)
China	2,282	10,531
Japan	1,167	7,482
USA	3,766	5,640
Iceland	1,469	5,603
Turkey	820	4,377
New Zealand	308	1,967
Georgia	250	1,752
Russia	308	1,707
France	326	1,360
Sweden	377	1,147
Hungary	473	1,135
Mexico	164	1,089
Italy	326	1,048
Rumania	152	797
Switzerland	547	663

(therapeutic baths), industrial processes, and electricity generation. The main types of direct use are bathing, swimming and balneology (42%), space heating (35%), greenhouses (9%), fish farming (6%), and industry (6%) (Fridleifsson 2001).

One of the most abundant energy sources on the surface of the Earth is sunlight. Today, solar energy has a tiny contribution in the world total primary energy supply, less than 1.0 (Ramachandra 2007). The potential of solar energy – passive solar heat, collectors for, e.g., hot water, and photovoltaic (PV) power – is tremendous.

Following the oil crises of the 1970s, energy experts began to explore whether solar-based power generation held potential as an alternative to petroleum-based fuels. Development of solar power has progressed considerably since then, yet its record of performance has been mixed, and it has not come into widespread use in either industrialized or developing countries.

PV systems, other than solar home heating systems, are used for communication, water pumping for drinking and irrigation, and electricity generation. The total installed capacity of such systems is estimated at about 1,000 kW. A solar home heating system is a solar PV system with a maximum capacity of 40 W. These systems are installed and managed by a household or a small community (Garg and Datta 1998).

Like wind power markets, PV markets have seen rapid growth and costs have fallen dramatically. The total installed capacity of such systems is estimated at about 1,000 kW. Solar PV is growing fast; the PV and grid-connected, wind-installed capacities are growing at a rate of 30% a year (Demirbas 2005).

Wind energy is a significant resource; it is safe, clean, and abundant. Wind energy is an indigenous supply permanently available in virtually every nation in the world. Using the wind to produce electricity by turning blades on a wind turbine is known as wind energy or wind power. More recently large wind turbines have been designed that are used to generate electricity. Wind as a source of energy is nonpolluting and freely available in many areas. As wind turbines are becoming more efficient, the cost of the electricity they generate is falling.

Wind power in coastal and other windy regions is promising as well. Today there are wind farms around the world. Production of wind-generated electricity has risen from practically zero in the early 1980s to more than 7.5 TWh per year in 1995. Cumulative generating capacity worldwide topped 6,500 MW in late 1997 (Demirbas 2005). Figure 1.1 shows the growth in world wind turbine installed capacity. Globally, wind power generation more than quadrupled between 1999 and 2005.

Wind energy is abundant, renewable, widely distributed, and clean and mitigates the greenhouse effect if it is used to replace fossil-fuel-derived electricity. Wind energy has limitations based on geography and climate, plus there may be political or environmental problems (e.g., dead birds) associated with installing turbines (Garg and Datta 1998). On the other hand, wind can contribute to air pollution by degrading and distributing pieces of pollutants such as waste paper, straw, etc.

Worldwide developments in the field of energy supply, following the oil crises of the 1970s and 2004, are showing the way to more serious decisions regarding sustainability in strategic energy planning, improvements in energy efficiency, and the rational use of energy. Renewable energy sources are increasingly becoming a key factor in this line of thought.

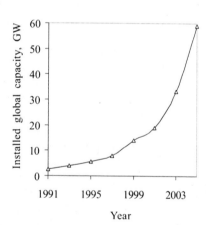

Figure 1.1 Growth in world wind turbine installed capacity

1.3 Present Energy Situation

1.3.1 Energy Production and Future Energy Scenarios

The world energy consumption pattern is also increasing, as shown in the Figure 1.2. Energy consumption has been increasing and will triple in a period of 50 years by

1.3 Present Energy Situation

Figure 1.2 World energy consumption pattern

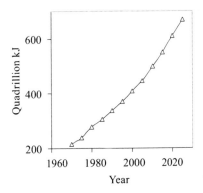

2025 as seen from Figure 1.2. The world's population will increase from 6 billion to 11 billion this century, life expectancy has doubled in the last two centuries, and energy requirements have increased 35 times in the same period. The main drivers of the search for alternative sources of energy are population growth, economics, technology, and agriculture.

1.3.2 Future Energy Scenarios

Fossil fuels still represent over 80% of total energy supplies in the world today, but the trend toward new energy sources is clear thanks to recent technological developments.

Oil is the fossil fuel that is most in danger of becoming scarce. The Middle East is the dominant oil province of the world, controlling 63% of the global reserves. Figure 1.3 shows global oil production scenarios based on today's production. A peak

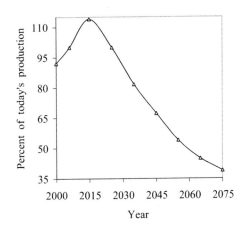

Figure 1.3 Global oil production scenarios based on current production

in global oil production may occur between 2015 and 2030. Countries in the Middle East and the Russian Federation hold 70% of the world's dwindling reserves of oil and gas. The geographical distribution of energy reserves and resources is important.

The peak of world gas production may not occur until 2025, but two things are certain: we will have even less warning than we had for Peak Oil, and the subsequent decline rates may be shockingly high. The NG model shows a plateau in production between 2025 and 2030. This is followed by a rapid increase in decline to 8% per year by 2050, remaining at a constant 8% per year for the following 50 years. This gives the production curve shown in Figure 1.4 (Kirtay 2009).

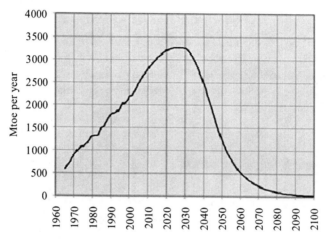

Figure 1.4 Global natural gas production, 1965–2100

Coal is the ugly stepsister of fossil fuels. Most coal today is used to generate electricity. As economies grow, so does their demand for electricity, and if electricity is used to replace some of the energy lost due to the decline of oil and natural gas, this will put yet more upward pressure on the demand for coal. Just as we saw with oil and gas, coal will exhibit an energy peak and decline. One factor in this is that we have in the past concentrated on finding and using the highest grade of coal, anthracite. Much of what remains consists of lower-grade bituminous and lignite. These grades of coal produce less energy when burned and require the mining of ever more coal to get the same amount of energy. Figure 1.5 shows global coal production, 1965 to 2100 (Kirtay 2009).

Renewable energy is a promising alternative solution because it is clean and environmentally safe. Renewables also produce lower or negligible levels of greenhouse gases and other pollutants when compared with the fossil energy sources they replace. Table 1.8 shows the global renewable energy scenario by 2040. Approximately half of the global energy supply will come from renewables in 2040 according to the European Renewable Energy Council (EREC 2006). Between 2001

1.3 Present Energy Situation

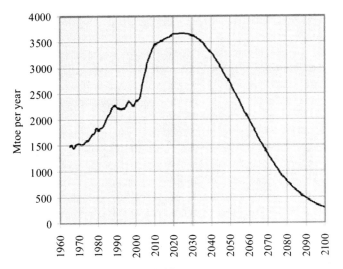

Figure 1.5 Global coal production, 1965 to 2100

Table 1.8 Global renewable energy scenario by 2040

	2001	2010	2020	2030	2040
Total consumption (Million ton oil equivalent)	10,038	10,549	11,425	12,352	13,310
Biomass	1,080	1,313	1,791	2,483	3,271
Large hydro	22.7	266	309	341	358
Geothermal	43.2	86	186	333	493
Small hydro	9.5	19	49	106	189
Wind	4.7	44	266	542	688
Solar thermal	4.1	15	66	244	480
Photovoltaic	0.2	2	24	221	784
Solar thermal electricity	0.1	0.4	3	16	68
Marine (tidal/wave/ocean)	0.05	0.1	0.4	3	20
Total renewable energy sources	1,365.5	1,745.5	2,694.4	4,289	6,351
Contribution of renewable energy sources (%)	13.6	16.6	23.6	34.7	47.7

and 2040 the most significant developments in renewable energy production will be observed in photovoltaics (from 0.2 to 784 Mtoe) and wind energy (from 4.7 to 688 Mtoe).

Biomass is the most used renewable energy source now and will be in the foreseeable future. The potential of sustainable large hydro is quite limited in some regions of the world. The potential for small hydro (<10 MW) power is still significant and will become increasingly significant in the future. Wind energy usage has grown by more than 30% annually in recent years, and this trend will likely increase in the future. Photovoltaics has already experienced impressive annual growth rates of more that 30% in recent years, and this trend promises to speed up in the future.

Geothermal and solar thermal sources will play more important roles in the future energy mix.

The use of fossil fuels as the primary energy source has led to a serious energy crisis and environmental pollution on a global scale. In order to mitigate environmental problems, the cost of renewable energy can be made competitive with fossil-fuel or nuclear energy by rapid technological developments in solar and wind energies. In order to mitigate environmental problems, renewable energy especially wind and solar energies at competitive costs resulting from the fast technological development. The limitations of solar power are site specific, intermittent, and, thus, not reliable for instantaneous supply. Using batteries to store any energy surplus for later consumption can resolve the time mismatch between energy supply and demand. The shortcomings of battery storage are low storage capacity, short equipment life, and considerable solid and chemical wastes generated. A system consisting of photovoltaic (PV) panels coupled with electrolyzers is a promising design for producing hydrogen (Ni *et al.* 2006).

A detailed analysis of the technical, economic, and regulatory issues of wind power can be found in the European Wind Energy Association (EWEA) report: "Large scale integration of wind energy in the European power supply: Analysis, issues and recommendations," published in December 2005. In 2005, worldwide capacity of wind-powered generators was 58,982 MW; although it currently produces less than 1% of worldwide electricity use, it accounts for 23% of electricity use in Denmark, 4.3% in Germany, and approx. 8% in Spain. Globally, wind power generation more than quadrupled between 1999 and 2005 according to the EWEA (2005).

Figure 1.6 shows the growth scenarios for global installed wind power (IEA 2004). In 2004, the International Energy Agency (IEA) Reference Scenario projections for wind energy were updated to 66 GW in 2010, 131 GW in 2020, and 170 GW in 2030. The IEA advanced strong growth scenario projected a wind energy market of 82 GW in 2010, 165 GW in 2020, and 250 GW in 2030.

Geothermal energy can be utilized in various forms such as electricity generation, direct use, space heating, heat pumps, greenhouse heating, and industrial heating. Electricity generation is improving faster in geothermal-energy-rich countries. As an energy source, geothermal energy has come of age. Utilization has increased rapidly during the last three decades.

Geothermal energy for electricity generation has been produced commercially since 1913, and for four decades on the scale of hundreds of megawatts both for electricity generation and direct use. In Tuscany, Italy, a geothermal plant has been operating since the early 1900s. There are also geothermal power stations in the USA, New Zealand, and Iceland. In Southampton (UK) there is a district heating scheme based on geothermal energy. A hot water is pumped up from about 1,800 m below ground. Use of geothermal energy has increased rapidly during the last three decades. In 2000, geothermal resources were identified in over 80 countries,and there are quantified records of geothermal utilization in 58 countries around the world (Fridleifsson 2001). Electricity is produced with geothermal steam in 21 countries spread over all continents. Low-temperature geothermal energy is

1.3 Present Energy Situation

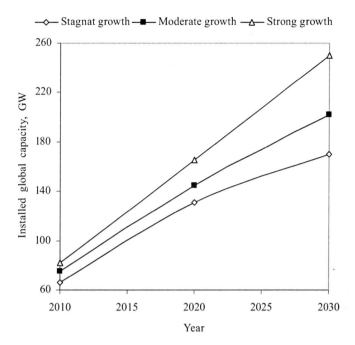

Figure 1.6 Growth scenarios for global installed wind power

exploited in many countries to generate heat, with an estimated capacity of approx. 10,000 MW thermal.

The world's total installed geothermal electric capacity was 7,304 MWe in 1996. In much of the world electricity from fossil-fuel-burning electricity plants can be provided at half the cost of new geothermal electricity. A comparison of renewable energy sources shows the current electrical energy cost to be US$ 0.02 to $ 0.10/kWh) for geothermal and hydro, US$ 0.05 to $ 0.13/kWh for wind, US$ 0.05 to $ 0.15/kWh for biomass, US$ 0.25 to $ 1.25/kWh for solar PV, and US$ 0.12 to $ 0.18/kWh for solar thermal electricity (Demirbas 2006).

Solar energy is defined as the radiant energy transmitted by the Sun and intercepted by the Earth. It is transmitted through space to Earth by electromagnetic radiation with wavelengths ranging between 0.20 and 15 µm. The availability of solar flux for terrestrial applications varies with season, time of day, location, and collecting surface orientation. In this chapter we shall treat these matters analytically (Kutz 2007).

One of the most abundant energy sources on the surface of the Earth is sunlight. Today, solar energy has a tiny contribution in the world total primary energy supply of less than 1%. PV systems, other than solar home heating systems, are used for communication, water pumping for drinking and irrigation, and electricity generation. The total installed capacity of such systems is estimated at about 1,000 kW. A solar home heating system is a solar PV system with a maximum capacity of

40 W. These systems are installed and managed by a household or a small community.

Like wind power markets, PV markets have seen rapid growth and costs have fallen dramatically. The total installed capacity of such systems is estimated at about 1,000 kW. PV installed capacities are growing at a rate of 30% a year. Solar PV systems hold great promise. One of the most significant developments in renewable energy production is observed in PVs. According to the European Wind Energy Association, PV will eventually be the largest renewable electricity source with a production of 25.1% of global power generation in 2040 (EWEA 2005).

A solar thermal electricity power system is a device that uses solar radiation for the generation of electricity through solar thermal conversion. Figure 1.7 shows the plot for electricity costs from solar thermal power plants. Solar thermal electricity may be defined as the result of a process by which directly collected solar energy is converted into electricity through the use of some sort of heat-to-electricity conversion device. The last three decades have witnessed a trend in solar thermal electricity generation of increasing the concentration of sunlight. There are three main systems of solar thermal electricity: solar towers, dishes, and parabolic troughs. Solar thermal power stations based on parabolic and heliostat trough concentrating collectors may soon become a competitive option on the world's electricity market. Table 1.9 shows the economics and emissions of conventional technologies compared with solar power generation.

PV systems convert sunlight directly into electricity. They work any time the sun is shining, but more electricity is produced when the sunlight is more intense and

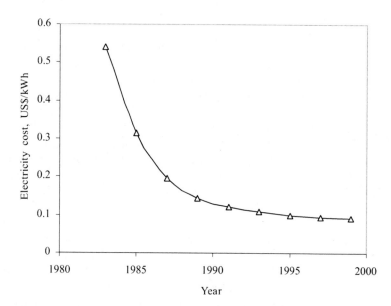

Figure 1.7 Electricity costs from solar thermal power plants

1.3 Present Energy Situation

Table 1.9 Economics and emissions of conventional technologies compared with solar power generation

Electricity generation technology	Carbon emissions g C/kWh	Generation costs US$/kWh
Solar thermal and solar PV systems	0	0.09–0.40
Pulverized coal–natural gas turbine	100–230	0.05–0.07

strikes the PV modules directly. The basic building block of PV technology is the solar "cell." Multiple PV cells are connected to form a PV "module," and the smallest PV component is sold commercially. Modules range in power output from about 10 W to 300 W. A PV system connected to the utility grid has the following components: (1) one or more PV modules connected to an inverter, (2) the inverter, which converts the system's direct-current (DC) electricity to alternating current (AC) and (3) batteries (optional) to provide energy storage or backup power in case of a power interruption or outage on the grid. Figure 1.8 shows a field of solar panels.

Figure 1.8 Solar panel field

Small, single-PV-panel systems with built-in inverters that produce about 75 W of electricity may cost around $900 installed, or $12 per watt. A 2-kW system that meets nearly all the needs of a very energy-efficient home could cost $16,000 to $20,000 installed, or $8 to $10 per watt. At the high end, a 5-kW system that completely meets the energy needs of many conventional homes can cost $30,000 to $40,000 installed, or $6 to $8 per watt. Figure 1.9 shows the schematic of a fuel cell.

Figure 1.10 shows the configuration of PV system with AC appliances (e.g., household lighting, heating, refrigeration, television, or video).

The term biofuel refers to liquid or gaseous fuels for the transport sector that are predominantly produced from biomass. Biofuels are mainly bioethanol, biomethanol, biodiesel, biohydrogen, and biogas. There are several reasons for biofuels to be considered as relevant technologies by both developing and industrialized countries. They include energy security reasons, environmental concerns, foreign exchange savings, and socioeconomic issues related to the rural sector.

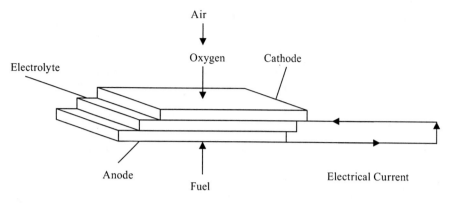

Figure 1.9 Fuel cell

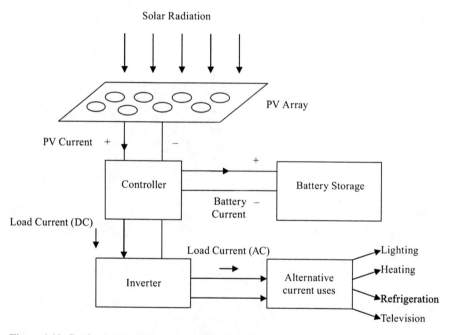

Figure 1.10 Configuration of PV system with AC appliances

Biofuels are important because they replace petroleum fuels. Biofuels are generally considered as addressing many of today's energy-related concerns, including sustainability, reduction of greenhouse gas emissions, regional development, social structure and agriculture, and security of supply (Reijnders 2006).

Biomass provides a number of local environmental gains. Energy forestry crops have a much greater diversity of wildlife and flora than the alternative land use, which is arable or pasture land. In industrialized countries, the main biomass pro-

1.3 Present Energy Situation

cesses utilized in the future will be expected to be direct combustion of residues and wastes for electricity generation, bioethanol and biodiesel as liquid fuels, and combined heat and power production from energy crops. The future of biomass electricity generation lies in biomass integrated gasification/gas turbine technology, which offers high energy-conversion efficiencies. Biomass will compete favorably with fossil mass for niches in the chemical feedstock industry. Biomass is a renewable, flexible, and adaptable resource. Crops can be grown to satisfy changing end use needs.

In the future, biomass will have the potential to provide a cost-effective and sustainable supply of energy, while at the same time aiding countries in meeting their greenhouse-gas-reduction targets. By the year 2050, it is estimated that 90% of the world's population will live in developing countries.

According to the IEA, scenarios developed for the USA and the EU indicate that near-term targets of up to 6% displacement of petroleum fuels with biofuels appear feasible using conventional biofuels, given available cropland. A 5% displacement of gasoline in the EU requires about 5% of available cropland to produce ethanol, while in the USA 8% is required. A 5% displacement of diesel requires 13% of US cropland, 15% in the EU (IEA 2006).

The recent commitment by the US government to increase bioenergy threefold in 10 years has added impetus to the search for viable biofuels. The advantages of biofuels are that they (a) are easily available from common biomass sources, (b) indicate a carbon dioxide cycle in combustion, (c) have a very environmentally friendly potential, (d) benefit the environment, economy, and consumers, and (e) are biodegradable and contribute to sustainability (IEA 2004).

Dwindling fossil fuel sources and the increasing dependency of the USA on imported crude oil have led to a major interest in expanding the use of bioenergy. In addition to increased interest by the US government, the EU has also adopted a proposal for a directive on the promotion of the use of biofuels with measures ensuring that biofuels account for at least 2% of the market for gasoline and diesel sold as transport fuel by the end of 2005, increasing in stages to a minimum of 5.75% by the end of 2010 (Puppan 2002). Bioethanol is a fuel derived from renewable sources of feedstock, typically from plants such as wheat, sugar beet, corn, straw, and wood. Bioethanol is a petrol additive/substitute. Biodiesel is superior to diesel fuel in terms of sulfur content, flash point, aromatic content, and biodegradability (Hansen *et al.* 2005).

If biodiesel is used for engine fuel, this would in turn benefit the environment and local populations. The benefits of biofuels over traditional fuels include greater energy security, reduced environmental impact, foreign exchange savings, and socioeconomic issues related to the rural sector.

Figure 1.11 shows the rate of consumption of alternative fuels compared to total automotive fuel consumption worldwide as a futuristic view (Demirbas 2006). Hydrogen is currently more expensive than conventional energy sources. There are different technologies presently being applied to produce hydrogen economically from biomass. Biohydrogen technology will play a major role in the future because it will allow for the use of renewable sources of energy (Nath and Das 2003).

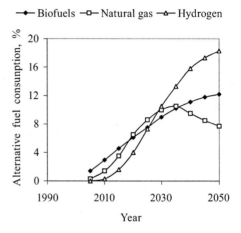

Figure 1.11 Rate of consumption of alternative fuels compared to total automotive fuel consumption worldwide

Biofuels include bioethanol, biobutanol, biodiesel, vegetable oils, biomethanol, pyrolysis oils, biogas, and biohydrogen. There are two global biomass-based liquid transportation fuels that might replace gasoline and diesel fuel. These are bioethanol and biodiesel. World production of biofuel was about 68 billion L in 2007 (59 billion L bioethanol and 9 million L biodiesel). The primary feedstocks of bioethanol are sugar cane and corn.

Bioethanol is a gasoline additive/substitute. Bioethanol is by far the most widely used biofuel for transportation worldwide. Global bioethanol production more than doubled between 2000 and 2005. About 60% of global bioethanol production comes from sugar cane and 40% from other crops. Biodiesel refers to a diesel-equivalent mono-alkyl-ester-based oxygenated fuel. Biodiesel production using inedible vegetable oil, waste oil, and grease has become more attractive recently. The economic performance of a biodiesel plant can be determined once certain factors are identified, such as plant capacity, process technology, raw material cost, and chemical costs. Even with today's high oil prices, biofuels cost more than conventional fuels. The central policy of biofuels grows out of concerns related to job creation, greater efficiency in the general business environment, and protection of the environment.

The biggest difference between biofuels and petroleum feedstocks is oxygen content. Biofuels have oxygen levels from 10 to 45%, while petroleum has essentially none, making the chemical properties of biofuels very different from those of petroleum. All have very low sulfur levels and many have low nitrogen levels. Biofuels are nonpolluting, locally available, accessible, sustainable, and reliable fuels obtained from renewable sources.

Sustainable biofuel production mainly depends on the productivity and the planting options. A sustainable use of biomass for bioenergy production is expected to reduce environmental contamination. Achieving a solution to environmental problems requires long-term policies for sustainable development. In this view, renewable energy sources appear to be one of the most efficient and effective solutions. Sustainability of renewable energy systems must support both human and ecosys-

1.3 Present Energy Situation

tem health over the long term; goals on tolerable emissions should look well into the future. The sustainability of biofuels for energy use requires a high efficiency recycling of energy and low emissions of carbon compounds, NO_x, persistent organics and acidifying compounds, and heavy metals due to biomass combustion. Electricity generation from biofuels holds great promise in the near future. Electricity costs are in the 6 to 8 c/kWh range.

Liquid biofuels for transportation have recently attracted consierable attention in various countries because of their renewability, sustainability, biodegradability, and common availability, as well as for their potential role in regional development, creation of rural manufacturing jobs, and reduction of greenhouse gas emissions. Table 1.10 shows the availability of modern transportation fuels. The advantage of biofuels in this regard is that they are derived from natural products. Policy drivers for renewable liquid biofuels have attracted particularly high levels of assistance in some countries given their promise of benefits in several areas of interest to governments, including agricultural production, greenhouse gas emissions, energy security, trade balances, rural development, and economic opportunities for developing countries. The EU ranks third in biofuel production worldwide, behind Brazil and the USA. In Europe, Germany is the largest produce of biofuels, while France is the second largest. Most biofuels in commercial production in Europe today are based on sugar beet, wheat, and rapeseed, which are converted into bioethanol/ETBE and biodiesel.

Table 1.10 Availability of modern transportation fuels

Fuel type	Availability	
	Current	Future
Gasoline	Excellent	Moderate-poor
Bioethanol	Moderate	Excellent
Biodiesel	Moderate	Excellent
Compressed natural gas (CNG)	Excellent	Moderate
Hydrogen for fuel cells	Poor	Excellent

Ethanol has attracted considerable interest in both industry and research as a plausible renewable energy source in the future. Bioethanol is a gasoline additive/substitute. Biomass wastes that with a high hydrocarbon content, such as sugar cane, sugar beets, corn, and molasses, can be good sources of bioethanol. Bioethanol is an alternative fuel that is produced almost entirely from food crops. Bioethanol represents an important, renewable liquid fuel for motor vehicles. It is derived from alcoholic fermentation of sucrose or simple sugars, which are produced from biomass by hydrolysis. In order to produce bioethanol from cellulosic biomass, a pretreatment process is used to reduce the sample size, break down the hemicelluloses into sugars, and open up the structure of the cellulose component. The cellulose portion is hydrolyzed by acids or enzymes into glucose sugar that is fermented into bioethanol.

The sugars from hemicelluloses are also fermented bioethanol. Producing and using bioethanol as a transportation fuel can help reduce carbon dioxide buildup in two important ways: by displacing the use of fossil fuels and by recycling the carbon dioxide that is released when combusted as fuel. The use of ethanol-blended fuel for automobiles can significantly reduce petroleum use and greenhouse gas exhaust emissions. An important advantage of crop-based ethanol is its greenhouse benefits. Corn stover consists of the stalks, leaves, cobs, and husk. It is possible to convert corn stover economically into bioethanol.

Ethanol can be produced in a culture medium. For this purpose, an alginate-loofa matrix was developed as a cell carrier for ethanol fermentation owing to its porous structure and strong fibrous nature. The matrix was effective for cell immobilization and had good mechanical strength and stability for long-term use (Phisalaphong *et al.* 2007).

Partly due to the oil crises, biomass-derived syngas (biosyngas) has become an important part of alternative energy since the 1980s. Once clean biosyngas is available, the known process technology can be used to produce biomethanol, Fischer–Tropsch diesel oil, and hydrogen. Methanol can be produced from hydrogen-carbon oxide mixtures by means of the catalytic reaction of carbon monoxide and some carbon dioxide with hydrogen. Biosynthesis gas (biosyngas) is a gas rich in CO and H_2 obtained by gasification of biomass. The mixture of gases from organic waste materials is converted into methanol in a conventional steam-reforming/water-gas shift reaction followed by high-pressure catalytic methanol synthesis.

Biodiesel is known as a monoalkyl; examples include methyl and ethyl, which are esters of fatty acids. Biodiesel is produced from triglycerides by transesterification (Demirbas 2003). Environmental and political concerns are generating a growing interest in alternative engine fuels such as biodiesel. Biodiesel is the best candidate for diesel fuels in diesel engines. Biodiesel refers to a diesel-equivalent processed fuel derived from biological sources. Biodiesel is the name given to a variety of ester-based oxygenated fuel from renewable biological sources. It can be made from processed organic oils and fats. Biodiesel production using inedible vegetable oils, waste oil, and grease has also become more attractive recently. Biodiesels will play an important role in meeting future fuel requirements in view of their lower toxicity and have an edge over conventional diesel as they are obtained from renewable sources (Sastry *et al.* 2006).

In one study, cottonseed methyl ester was used in a four-stroke, single-cylinder, air-cooled diesel engine as an alternative fuel. Engine tests were carried out at full load and at different speeds. The engine torque and power of cottonseed oil methyl ester were found to be lower than that of diesel fuel in the range of 3 to 9% and specific fuel consumption was higher than that of diesel fuel by approx. 8 to 10%. CO_2, CO, and NO_x emissions of cottonseed methyl ester were lower than that of diesel fuel (Yucesu and Ilkilic 2006).

In general, the physical and chemical properties and the performance of the cottonseed oil methyl ester were comparable to those of diesel fuel (Ilkilic and Yucesu 2008). The effects of cottonseed oil methyl ester and diesel fuel on a direct-injected, four-stroke, single-cylinder, air-cooled diesel engine performance and exhaust emis-

1.3 Present Energy Situation

sions were investigated. Test quantities of cottonseed oil methyl ester of renewable fuels were processed and characterized, and performance and exhaust gas emissions were tested in various injection pressures. In order to determine emission and performance characteristics, the engine was tested with a full load and at various injection pressures and constant speed. The results showed that engine performance using cottonseed oil methyl ester fuel differed little from engine performance and torque with diesel fuel. As for the emissions, there was an approx. 30% reduction in CO and approx. 25% reduction in NO_x (Yucesu and Ilkilic 2006).

An engine performance test using sunflower methyl esters exhibited characteristics very similar to regular diesel. The test values obtained from a 2.5-L, 4-cylinder Peugeot XD3p157 engine showed that torque values obtained by the two types of fuels are 5 to 10% in favor of regular diesel. Specific fuel consumption, however, is better with biodiesel. This means that a better combustion characteristic is achieved with biodiesel, which compensates for its lower calorific value. Soot emissions are slightly less with biodiesel, as expected, due to the improvement in specific fuel consumption (Kaplan *et al.* 2006). The physical and chemical properties of methyl ester of waste cooking oil were determined in the laboratory. The methyl ester was tested in a diesel engine with turbocharged, four cylinders, and direct injection. Obtained results were compared with No. 2 diesel fuel (Utlu 2007).

A new lipase immobilization method, textile cloth immobilization, was developed for the conversion of soybean oil into biodiesel. Immobilized Candida lipase sp. 99–125 was applied as the enzyme catalyst. The effect of flow rate of the reaction liquid, solvents, reaction time, and water content on the biodiesel yield was investigated. The test results indicated that the maximum yield of biodiesel of 92% was obtained with hexane as the solvent, water content of 20 wt%, and reaction time of 24 h (Lv *et al.* 2008).

The dynamic transesterification reaction of peanut oil in supercritical methanol medium was investigated. The reaction temperature and pressure were in the range of 523 to 583 K and 10.0 to 16.0 MPa, respectively. The molar ratio of peanut oil to methanol was 1:30. It was found that the yield of methyl esters was higher than 90% in the supercritical methanol. The apparent reaction order and activation energy of transesterification was 1.5 and 7.472 kJ/mol, respectively. In this method, the reaction time was shorter and the processing was simpler than that of the common acid catalysis transesterification (Cheng *et al.* 2008).

The existing biodiesel production process is neither completely "green" nor renewable because it utilizes fossil fuels, mainly natural gas as an input for methanol production. Also the catalysts currently in use are highly caustic and toxic. To overcome the limitation of the existing process, a new method was proposed that used waste vegetable oil and nonedible plant oils as biodiesel feedstock and nontoxic, inexpensive, and natural catalysts. The economic benefit of the proposed method was also discussed. The new method will render the biodiesel production process truly green (Chhetri and Islam 2008).

A four-stroke, three-cylinder, 30-kW TUMOSAN (Turkish Motor Industry and Trade) diesel engine was used for experimentation with biodiesel. The kinematic viscosity, density, flash point, cloud point, pour point, freezing point, and copper strip corrosion values of all biodiesel fuels stayed within the limit values described by DIN-TSE EN 14214. They can readily be used without any need for modification to the engine. Even though specific fuel consumption for biodiesel fuels tended to be higher than that for normal diesel fuel, the exhaust smokiness values of biodiesel fuels were considerably lower than that for petroleum diesel fuel. On the other hand, there were no significant differences observed for torque, power, and exhaust smokiness (Oguz *et al.* 2007).

Problems to be studied include fuel storage stability, fuel solubility, and oxidative stability of recycled soybean-derived biodiesel. Unlike newly manufactured soy oils, it was found that this recycled soy oil was not stable in fuels. The question was what in the recycled oil led to the observed fuel degradation (Mushrush *et al.* 2007).

Oxidative and thermal degradation occurs on the double bonds of unsaturated aliphatic carbon chains in biolipids. Oxidation of biodiesel results in the formation of hydroperoxides. Oxidative and thermal instability is determined by the amount and configuration of the olefinic unsaturation on the fatty acid chains. The viscosity of biodiesel increases with an increase in the thermal degradation degree due to transisomer formation on double bonds. The decomposition of biodiesel and its corresponding fatty acids linearly increases from 293 to 625 K. The density of biodiesel fuels decreases linearly with temperature from 293 to 575 K. The combustion heat of biodiesel partially decreases with an increase in thermal degradation degree (Arisoy 2008).

The emission-forming gases, such as carbon dioxide and carbon monoxide from the combustion of biodiesel, generally are less than those of diesel fuel. Sulfur emissions are essentially eliminated with pure biodiesel. The exhaust emissions of sulfur oxides and sulfates from biodiesel are essentially eliminated compared to diesel. The smog-forming potential of biodiesel hydrocarbons is less than that of diesel fuel. The ozone-forming potential of speciated hydrocarbon emissions is 50% less than that measured for diesel fuel (Dincer 2008).

References

Arisoy, K. 2008. Oxidative and thermal instability of biodiesel. Energy Sources Part A 30:1516–1522.

Cheng, J., Li, Y., He, S., Shen, W., Liu, Y., Song, Y. 2008. Reaction kinetics of transesterification between vegetable oil and methanol under supercritical conditions. Energy Sources Part A 30:681–688.

Chhetri, A. B., Islam, M. R. 2008. Towards producing a truly green biodiesel. Energy Sources Part A 30:754–764.

Demirbas, A. 2003. Biodiesel fuels from vegetable oils via catalytic and non-catalytic supercritical alcohol transesterifications and other methods: a survey. Energy Convers Manage 44:2093–2109.

Demirbas, A. 2005. Potential applications of renewable energy sources, biomass combustion problems in boiler power systems and combustion related environmental issues. Prog Energy Combus Sci 31:171–192.

Demirbas, A. 2006. Biofuel based cogenerative energy conversion systems. Energy Sources Part A 28:1509–1518.

References

Dincer, K. 2008. Lower emissions from biodiesel combustion. Energy Sources Part A 30:963–968.

EREC. 2006. Renewable energy scenario by 2040, EREC Statistics, Brussels.

EWEA. 2005. Report: Large scale integration of wind energy in the European power supply: analysis, issues and recommendations, Paris.

Fridleifsson, I. B. 2001. Geothermal energy for the benefit of the people. Renew Sustain Energy Rev 5:299–312.

Garg, H. P., Datta, G. 1998. Global status on renewable energy, in Solar Energy Heating and Cooling Methods in Building, International Workshop: Iran University of Science and Technology. 19–20 May.

Hall, D. O., Rosillo-Calle, F., de Groot, P. 1992. Biomass energy lessons from case studies in developing countries. Energy Policy 20:62–73.

Hansen, A. C., Zhang, Q., Lyne, P. W. L. 2005. Ethanol–diesel fuel blends–a review. Biores Technol 96:277–285.

IEA. 2004. Biofuels for transport: an international perspective. 9, rue de la Fédération, 75739 Paris cedex 15, France (available from: www.iea.org).

IEA. 2006. Reference scenario projections. 75739 Paris cedex 15, France.

IEA. 2007. Key world energy statistics. Paris. http://www.iea.org/Textbase/nppdf/free/2007/key_stats_2007.pdf.

Jean-Baptiste, P., Ducroux, R. 2003. Energy policy and climate change. Energy Policy 31:155–166.

Kaplan, C., Arslan, R., Surmen, A. 2006. Performance characteristics of sunflower methyl esters as biodiesel. Energy Sources Part A 28:751–755.

Kirtay, E. 2009. World's population and energy demand. Soc Politic Econ Cultur Res 1:145–169.

Kutz, M. (ed.). 2007. Environmentally Conscious Alternative Energy Production. Wiley, Hoboken, NJ.

Lv, P., Wang, X., Yuan, Z., Tan, T. 2008. Conversion of soybean oil to biodiesel fuel with immobilized Candida lipase on textile cloth. Energy Sources Part A 30:872–879.

Mushrush, G. W., Wynne, J. H., Lloyd, C. T., Willauer, H. D., Hughes, J. M. 2007. Instability reactions and recycled soybean-derived biodiesel fuel liquids. Energy Sources Part A 29:491–497.

Nath, K., Das, D. 2003. Hydrogen from biomass. Curr Sci 85:265–271.

Ni, M., Leung, M. K. H., Sumathy, K., Leung, D. Y. C. 2006. Potential of renewable hydrogen production for energy supply in HongKong. Int J Hydrogen Energy 31:1401–1412.

Oguz, H., Ogut, H., Eryilmaz, T. 2007. Investigation of biodiesel production, quality and performance in Turkey. Energy Sources Part A 29:1529–1535.

Phisalaphong, M., Budiraharjo, R., Bangrak, P., Mongkolkajit, J., Limtong, S. 2007. Alginate-loofa as carrier matrix for ethanol production. J Biosci Bioeng 104:214–217.

Puppan, D. 2002. Environmental evaluation of biofuels. 2002. Periodica Polytechnica Ser Soc Man Sci 10:95–116.

Ramachandra, T. V. 2007. Solar energy potential assessment using GIS. Energy Educ Sci Technol 18:101–114.

Ramage, J., Scurlock, J. 1996. Biomass, in Renewable Energy-Power for a Sustainable Future, ed. G. Boyle, Oxford University Press, Oxford.

Reijnders, L. 2006. Conditions for the sustainability of biomass based fuel use. Energy Policy 34:863–876.

Sastry, G. S. R., Murthy, A. S. R. K., Prasad, P. R., Bhuvaneswari, K., Ravi, P. V. 2006. Identification and determination of bio-diesel in diesel. Energy Sources Part A 28:1337–342.

Schobert, H. H., Song, C. 2002. Chemicals and materials from coal in the 21st century. Fuel 81:15–32.

UNDP. 2000. World Energy Assessment. Energy and the challenge of sustainability. United Nations, New York.

Utlu, Z. 2007. Evaluation of biodiesel fuel obtained from waste cooking oil. Energy Sources Part A 29:1295–1304.

Yucesu, H. S., Ilkilic, C. 2006. Effect of cotton seed oil methyl ester on the performance and exhaust emission of a diesel engine. Energy Sources Part A 28:389–398.

Chapter 2
Green Energy Facilities

2.1 Introduction

Modern bioenergy involves commercial energy production from biomass for industry, power generation, or transport fuels. Bioenergy is an all-inclusive term for all forms of biomass and biofuels. Green energy is an alternate term for renewable energy used to indicate that the energy is generated from sources considered environmentally friendly. Various biomasses such as wood, straw, and even household wastes may be economically converted into bioethanol. Bioethanol is derived from the alcoholic fermentation of sucrose or simple sugars, which are produced from biomass by a hydrolysis process. Biodiesel is an environmentally friendly alternative liquid fuel that can be used in any diesel engine without modification.

The importance of biomass in different world regions is given in Table 2.1. For large portions of the rural populations of developing countries, and for the poorest sections of urban populations, biomass is often the only available and affordable

Table 2.1 Importance of biomass in different world regions

Region	Share of biomass in final energy consumption
Africa (average)	62.0
Burundi	93.8
Ethiopia	85.6
Kenya	69.6
Somalia	86.5
Sudan	83.7
Uganda	94.6
South Asia (average)	56.3
East Asia (average)	25.1
China	23.5
Latin America (average)	18.2
Europe (average)	03.5
North America (average)	02.7
Middle East (average)	0.3

A. Demirbas, M. Fatih Demirbas, *Algae Energy*
DOI 10.1007/978-1-84996-050-2, © Springer 2010

source of energy for basic needs such as cooking and heating. As shown in Table 2.1, the importance of biomass varies significantly across regions. In Europe, North America, and the Middle East, the share of biomass averages 2 to 3% of total final energy consumption, whereas in Africa, Asia, and Latin America, which together account for three-quarters of the world's population, biomass provides a substantial share of the energy needs: a third on average, but as much as 80 to 90% in some of the poorest countries of Africa and Asia (e.g., Angola, Ethiopia, Mozambique, Tanzania, Democratic Republic of Congo, Nepal, and Myanmar).

Traditional biomass markets have been inefficient, but technological developments have reduced energy, emissions, and material flows through the system thus improving the efficiency of biomass energy systems. The energy market demands cost effectiveness, high efficiency, and reduced risk to future emission limits. Modern biomass conversion technologies offer (a) the potential for high yields, (b) economic fuel availability, (c) low adverse environmental impacts, and (d) suitability to modern energy systems. A number of systems that meet the aforementioned criteria for modernized biomass conversion can be identified (Larson 1993).

2.2 Biomass Feedstocks

The term biomass (Greek *bio*, life + *maza* or mass) refers to nonfossilized and biodegradable organic material originating from plants, animals, and microorganisms derived from biological sources. Biomass includes products, byproducts, residues and waste from agriculture, forestry, and related industries as well as the nonfossilized and biodegradable organic fractions of industrial and municipal solid wastes. The term biomass refers to wood, short-rotation woody crops, agricultural wastes, short-rotation herbaceous species, wood wastes, bagasse, industrial residues, waste paper, municipal solid waste, sawdust, biosolids, grass, waste from food processing, aquatic plants and algae animal wastes, and a host of other materials.

Biomass is organic material that has stored sunlight in the form of chemical energy. It is commonly recognized as an important renewable energy; during the growth of plants and trees, solar energy is stored as chemical energy via photosynthesis, which can be released via direct or indirect combustion.

Biomass is very important for implementing the Kyoto agreement to reduce carbon dioxide emissions by replacing fossil fuels. Surging energy requirements in parallel with technological developments have forced research and development activities to focus on new and biorenewable energy.

There are three ways to use biomass. It can be burned to produce heat and electricity, changed to gaslike fuels such as methane, hydrogen, and carbon monoxide, or changed to a liquid fuel. Liquid fuels, also called biofuels, include mainly two forms of alcohol: ethanol and methanol. Because biomass can be converted directly into a liquid fuel, it could some day supply much of our transportation fuel needs for cars, trucks, buses, airplanes, and trains. This is very important because nearly one-third of our nation's energy is now used for transportation.

2.3 Green Energy

Biomass feedstocks are marked by their tremendous diversity, which makes them rather difficult to characterize as a whole. Feedstocks that can be utilized with conversion processes are primarily the organic materials now being landfilled. These include forest product wastes, agricultural residues, organic fractions of municipal solid wastes, paper, cardboard, plastic, food waste, green waste, and other waste. Nonbiodegradable organic feedstocks, such as most plastics, are not convertible by biochemical processes. Biobased materials require pretreatment by chemical, physical, or biological means to open up the structure of the biomass. The general categories of biomass feedstock are shown in Table 2.2.

Table 2.2 Major categories of biomass feedstocks

Forest products	Wood; logging residues; trees, shrubs, and wood residues; sawdust, bark, etc.
Biorenewable wastes	Agricultural wastes, crop residues, mill wood wastes, urban wood wastes, urban organic wastes
Energy crops	Short-rotation woody crops, herbaceous woody crops, grasses, starch crops, sugar crops, forage crops, oilseed crops, switchgrass, miscanthus
Aquatic plants	Algae, Water weed, water hyacinth, reed and rushes
Food crops	Grains, oil crops
Sugar crops	Sugar cane, sugar beets, molasses, sorghum
Landfill	Hazardous waste, nonhazardous waste, inert waste, liquid waste
Organic wastes	Manucipal solid waste, industrial organic wastes, municipal sewage and sludges
Algae	Prokaryotic algae, eukaryotic algae, kelps
Mosses	Bryophyta, polytrichales
Lichens	Crustose lichens, foliose lichens, fruticose lichen

2.3 Green Energy

Green energy is an alternate term for renewable energy used to indicate that the energy is generated from sources considered environmentally friendly (e.g., hydro, solar, biomass, or wind). Green power is sometimes used in reference to electricity generated from "green" sources. Green energy production is the principal contributor to the economic development of developing countries, whose economic development is based on agricultural production and where most people live in rural areas. Implementation of integrated community development programs is therefore very necessary.

Green power refers to electricity supplied from energy sources that are more readily renewable than traditional electrical power sources. Green power products, which can be derived from renewable energy sources, have become widespread in many electricity markets worldwide . The environmental advantages of green electricity seem clear. Market research indicates that there is a large potential market for green energy in Europe generally. Green power marketing has emerged in more than a dozen countries around the world.

Many green electricity products are based on renewable energy sources like wind, biomass, hydro, biogas, solar, and geothermal (Murphya and Niitsuma 1999). There has been interest in electricity from renewable sources known as green electricity or green pools as a special market (Elliott 1999). The term green energy is also used to describe green energy produced from cogeneration, energy from municipal waste, natural gas, and even conventional energy sources. The use of green energy sources like hydro, biomass, geothermal, and wind energy in electricity production reduces CO_2 emissions (Fridleifsson 2003). Emissions such as SO_2, CO_2, and NO_x are reduced considerably and the production and use of green electricity contributes to diminishing the greenhouse effect (Arkesteijn and Oerlemans 2005).

In general, a sustainable energy system includes energy efficiency, energy reliability, energy flexibility, energy development and continuity, combined heat and power (CHP) or cogeneration, fuel poverty, and environmental impacts. The environmental impacts of energy use are not new. For centuries, wood burning has contributed to the deforestation of many areas. On the other hand, the typical characteristics of a sustainable energy system can be derived from political definitions. A sustainable energy system can be defined also by comparing the performance of different energy systems in terms of sustainability indicators (Alanne and Sari 2006). Because, by definition, sustainable energy systems must support both human and ecosystem health over the long term, goals on tolerable emissions should look well into the future. They should also take into account the public's tendency to demand more (UNDP 2000).

2.4 Importance of Biomass

Biomass is organic material that has stored sunlight in the form of chemical energy. It is a general term for material derived from growing plants or from animal manure. Biomass includes wood and logging residues, agricultural crops and their waste byproducts, the organic portion of municipal solid waste, animal wastes, municipal biosolids (sewage), waste from food processing, and aquatic plants and algae (Hall *et al.* 1993). All biomass is produced by green plants converting sunlight into plant material through photosynthesis. Photosynthesis is a carbon fixation reaction achieved by the reduction of carbon dioxide. The fixation or reduction of carbon dioxide is a light-independent process. Although some of the steps in photosynthesis are still not completely understood, the overall photosynthetic equation has been known since the 1800s.

Biomass appears to be an attractive feedstock for three main reasons. First, it is a renewable resource that could be sustainably developed in the future. Second, it appears to have considerable environmentally friendly properties, resulting in no net releases of carbon dioxide and very low sulfur content. Third, it appears to have significant economic potential provided that fossil fuel prices increase in the future.

Agricultural residues such as straws, nut shells, fruit shells, fruit seeds, plant stalks and stovers, green leaves and molasses are potential renewable energy sources.

2.4 Importance of Biomass

Current disposal methods for these agricultural residues have caused widespread environmental concerns. For example, disposal of rice and wheat straw by open-field burning causes air pollution (Zhang and Zhang 1999).

Agricultural energy or green energy production is the principal contributor to the economic development of a developing country, whose economic development is based on agricultural production, and most people live in rural areas. Implementation of integrated community development programs is therefore very necessary. It is believed that integrated community development promotes the socioeconomic advancement of developing countries.

The biomass produced in Brazil results largely from an ethanol fuel production program started in 1975 from sugar cane crops grown specifically for fuel use, presently occupying 2.7 million hectares of land and employing about 350 distilleries. Ethanol currently provides over 40% of the fuel consumed by cars and light trucks. It is estimated to have saved Brazil over \$40 billion in oil imports, excluding the costs of the program. In 1999, carbon emissions were reduced by almost 13 megatons as a result of the program, and local emissions of lead, sulfur, and carbon monoxide have been greatly reduced. In addition, according to the United Nations Development Programme, ethanol production supports about 700,000 rural jobs (UNDP 2000).

Biomass has historically been a dispersed, labor-intensive, and land-intensive source of energy. Therefore, as industrial activity has increased in countries, more concentrated and convenient sources of energy have been substituted for biomass. Selection of a particular biomass for energy requirements is influenced by its availability, source and transportation costs, competing uses, and prevalent fossil fuel prices.

Biomass provides a clean, renewable energy source that could dramatically improve our environment, economy, and energy security. Biomass energy generates far less air emissions than fossil fuels, reduces the amount of waste sent to landfills, and decreases our reliance on foreign oil. Biomass energy also creates thousands of jobs and helps revitalize rural communities.

Energy can be obtained from direct combustion of biomass by burning of dry organic matter, such as woody scraps, grasses, and agricultural residues. Biomass can be thermochemically converted into liquid fuel and into gases, such as methane, carbon monoxide, or hydrogen, by pyrolysis. Bioethanol can be obtained from cellulosic biomass by fermenting and distilling sugar solutions. Vegetable oils such as soybean and canola oils can be chemically converted into liquid fuel known as biodiesel. These fuels can be used as diesel fuel and gasoline in conventional engines with little modification to the system. Certain organic compounds, specifically municipal biosolids (sewage) and animal wastes (manures), can be biochemically converted into methane by anaerobic digestion.

Certain transportation biofuels such as bioethanol, biodiesel, methyltetrahydrofuran, and dimethyl ether can be sustainably obtained from biomass. Biofuels can be used in pure form or blended with gasoline.

In industrialized countries, the main biomass processes utilized in the future are expected to be the direct combustion of residues and wastes for electricity generation, bioethanol and biodiesel as liquid fuels, and combined heat and power production from energy crops. The electricity is produced by the direct combustion of biomass and by advanced gasification and pyrolysis technologies that are almost ready for commercial-scale use.

Bioenergy, the energy from biomass, has been used for thousands of years, ever since people started burning wood to cook food or to keep warm, and today wood is still our largest biomass resource for bioenergy. Many countries in the developing world still use wood as their primary fuel (Demirbas and Demirbas 2007). In the future, biomass has the potential to provide a cost-effective and sustainable supply of energy, while at the same time aiding countries in meeting their greenhouse-gas-reduction targets. In the short to medium term, biomass is expected to dominate energy supply. For the generation of electricity and heat, with the use of advanced combustion technology, organic wastes can be used as biomass. Also a number of crops and crop residues may fit modern bioenergy chains (Pimentel *et al.* 1981; Haberl and Geissler 2000; Hoogwijk *et al.* 2003). Biomass is the most common form of carbonaceous materials, which are widely used in the third world.

Biomass gasification technologies make it possible to convert renewable biomass materials into clean fuel gases or synthetic gases. These gaseous products can be burned to generate heat or electricity, or they can potentially be used in the synthesis of liquid transportation fuels, hydrogen, or chemicals. Gasification offers a combination of flexibility, efficiency, and environmental acceptability that is essential in meeting future energy requirements. Biomass-fueled renewable distributed generation technologies in rural electrification provide no-regret options with significant CO_2 emission mitigation potential when operated under a net-metering scheme (Karki *et al.* 2008).

Cogeneration, in and of itself, is an example of pollution prevention. Cogenerators, by using excess heat, may enhance the efficiency of total energy use by up to 80% or more from the typical 33 to 38% efficiency of electricity-only generation. A typical cogeneration system consists of an engine, steam turbine, or combustion turbine that drives an electrical generator. Figure 2.1 shows a gas turbine topping cycle cogeneration system. The figure is a diagram of a gas turbine of the type used in cogeneration power plants ranging in size from about 500 kW to 50 MW. These turbines are similar to jet engines used in aircraft. In this size range, gas turbines are more efficient than any other type of fossil-fuel-fired power plants, reaching electrical efficiencies of up to 45% or even, with the addition of waste heat recovery (cogeneration), efficiencies of over 80%. When used for cogeneration, the hot exhaust gases from the gas turbine are passed through a heat exchanger where the heat is transferred to water in pipes producing steam. A waste heat exchanger recovers waste heat from the engine or exhaust gas to produce hot water or steam. Cogeneration produces a given amount of electric power and process heat with 15 to 35% less fuel than it takes to produce the electricity and process heat separately (Demirbas 2006c).

Figure 2.2 illustrates a comparison between energy inputs to separate and cogenerative generation systems. Traditional coal-, oil-, or natural-gas-fired thermal generating stations convert only about one-third of the initial energy contained within

2.4 Importance of Biomass

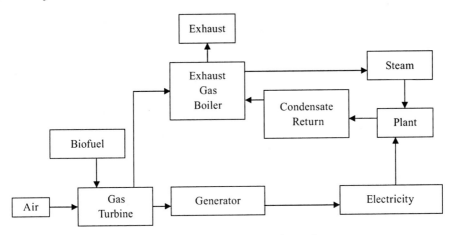

Figure 2.1 A representative cogeneration: a gas turbine topping cycle

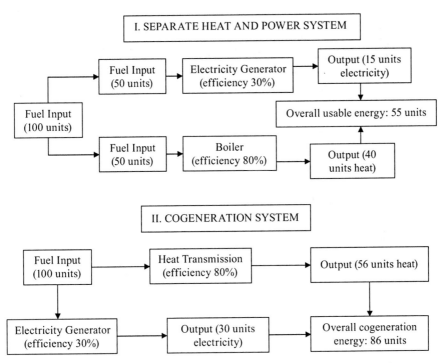

Figure 2.2 Comparison between energy inputs to separate heat and power system and cogeneration system

the fuel into useful electricity. The remainder of the energy is discarded as heat without serving any useful purpose. From 10 to 35% of primary energy use is wasted as conversion losses in power plants. Cogeneration can increase the efficiency of a fossil fuel from an average of 40 to over 80%. Figure 2.2 illustrates the increase in efficiency. This increase in efficiency can translate into lower costs and fewer emissions of pollutants than the conventional alternative of generating electricity and heat separately (Demirbas 2006c).

2.5 Production of Fuels and Chemicals from Biomass

Biomass provides a potential source of added value chemicals, such as reducing sugars, furfural, ethanol, and other products, by using biochemical or chemical and thermochemical processes. The gas phase of pyrolitic degradation products contains mostly carbon monoxide and carbon dioxide and minor proportions of hydrogen, methane, ethane, and propane. The liquid fraction consists mainly of water, with small proportions of acetaldehyde, propion aldehyde, butiraldehyde, acrolein, crotonaldehyde, furan, acetone, butanedione, and methanol. The sugars from the hemicelluloses are also fermented into bioethanol.

More recently, many attempts have been made to utilize biomass, such as wood, hazelnut shell, agricultural waste residues, waste paper and wheat straw, tea waste, and olive husk. The main biomass sources in use for energy production range from forest residues, agricultural residues, pulp and paper operation residues, animal waste, and landfill gas to energy crops.

Gasification followed by Fischer–Tropsch synthesis (FTS) is currently the most promising method for upgrading low-value coal and biomass to high-value liquid fuels and chemicals. The total biomass produced each year as waste material from agriculture and forest operations could be converted into roughly 40 billion gal/year of liquid fuels (roughly 25% of the current US gasoline usage).

Tijmensen *et al.* (2002) review the technical feasibility and economics of a biomass integrated gasification Fischer–Tropsch (BIG-FT) process and also point out the key R&D issues involved in the commercialization of this process. Boerrigter and den Uil (2002) give a similar review identifying a potential BIG-FT process configuration. The FTS for the production of liquid hydrocarbons from coal-based synthesis gas has been the subject of renewed interest for the conversion of coal and natural gas into liquid fuels (Jin and Datye 2000).

To produce biosyngas from a biomass fuel, the following procedures are necessary: (a) gasification of the biomass, (b) cleaning of the product gas, and (c) use of the synthesis gas to produce liquid fuel via FTS.

Hydrothermal liquefaction (HTL) or direct liquefaction is a promising technology to treat waste streams from various sources and produce valuable bioproducts such as biocrudes. A major problem with commercializing HTL processes for biomass conversion today is that it remains uneconomical when compared to the costs of diesel or gasoline production. High transportation costs of large quantities of biomass increase production costs, and a poor conversion efficiency, coupled with a lack of understanding of complex reaction mechanisms, inhibits the widespread commercial use of the process.

2.5 Production of Fuels and Chemicals from Biomass

In the HTU process, biomass is reacted in liquid water at elevated temperature and pressure. The phase equilibria in the HTU process are very complicated due to the presence of water, supercritical carbon dioxide, alcohols, and so-called biocrude. Biocrude is a mixture with a wide molecular weight distribution and consists of various kinds of molecules. Biocrude contains 10 to 13% oxygen and is upgraded by catalytic hydrodeoxygenation in a central facility. Preliminary process studies on the conversion of various biomass types into liquid fuels have indicated that HTU is more attractive than pyrolysis or gasification. In HTU the biomass, typically in a 25% slurry in water, is treated at temperatures of 575 to 700 K and 12 to 18 MPa pressures in the presence of liquid water for 5 to 20 min to yield a mixture of liquid biocrude, gas (mainly CO_2), and water. Subsequent processing may be able to upgrade the biocrude to useable biofuel. A large proportion of the oxygen is removed as carbon dioxide (Goudriaan and Peferoen 1990).

Biomass, such as wood, with a lower energy density is converted into biocrude with a higher energy density, organic compounds, mainly alcohols and acids, and gases, mainly CO_2. Water is also a byproduct. In the products, CO_2, the main component of the gas product, can be used to represent all gas produced, and methanol and ethanol represent organic compounds. The feedstocks, reaction conditions, and products for the HTU process are given in Table 2.3. In Table 2.3, the weight fraction of each component is assigned on the basis of the data of the vacuum flash of biocrude and the data of a pilot plant (Feng *et al.* 2004).

Table 2.3 Representatives for the products from the HTU process

Product	Component	Weight fraction (%)
Biocrude	Polycarbonates	47.5
	Methyl-*n*-propyl ether	2.5
Gas	Carbon dioxide	25.0
Organic compounds	Methanol	5.0
	Ethanol	3.5
Water	Water	16.5

One of the first HTL studies was conducted by Kranich (1984) using municipal waste materials (MWMs) as a source to produce oil. Three different types of materials from an MSW plant were used: primary sewage sludge, settled digester sludge, and digester effluent. Using a magnetically stirred batch autoclave with a hydrogen feed system, slurry feed device, a pressure and temperature recorder, and a wet-test meter for measuring gas product, Kranich processed the waste sources. The feedstock was first dried and then powdered. The wastes were also separated into different oil and water slurries and processed separately. Temperatures ranged from 570 to 720 K with pressures of up to 14 MPa. Retention times also varied between 20 and 90 min. Hydrogen was used as the reducing gas with initial pressures of up to 8.3 MPa. Three types of catalyst were studied: sodium carbonate, nickel carbonate, and sodium molybdate. The slurry feedstock was injected into the reactor through

38   2 Green Energy Facilities

a pressurized injector and the oil product was extracted by pentane and toluene. Results showed that organic conversion rates varied from 45 to 99% and oil production rates were reported from 35.0 to 63.3%. Gas products were found to contain H_2, CO_2, and C_1–C_4 hydrocarbons.

The experimental results showed no significant differences between the applications of the three different catalysts. Kranich recommended that the water slurry system was not feasible for scale-up, and considerations of a commercial scale process were confined to only the oil slurry system. It was also concluded that no further development work on hydroliquefaction of sewage sludge to oil was necessary. Kranich's recommendation did not hold, mainly due to increases in crude oil prices and the need to find new technologies for energy procurement, and thus many studies on liquefaction of sewage sludge have since been conducted. Research has indicated that liquefaction is a feasible method for the treatment of sewage sludge wastes and has a high oil-producing potential (Suzuki *et al.* 1986; Itoh *et al.* 1994; Inoue *et al.* 1997). Today HTL research is still being conducted with sewage sludge; however, focus has shifted to include many varieties of biomass materials. Table 2.4 shows feedstocks, reaction conditions, and the products for the HTU process.

Several technologies have been developed to convert biomass into a liquid biofuel with a higher heating value, such as gasification, fast pyrolysis, and HTU. In the HTU, the biomass is treated for 5 to 20 min with water under subcritical conditions (575 to 625 K, 10 to 18 MPa) to give a heavy organic liquid (biocrude) with a heating value of 30 to 35 MJ/kg. During this process, the oxygen content of the organic material is reduced from about 40% to between 10 and 15%. The removed oxygen ends up in CO_2, H_2O, and CO. After 1.6 s at 595 K and 25 MPa, 47% conversion of

Table 2.4 Feedstocks, reaction conditions, and products for HTU process

Biomass feedstocks	Wood and forest wastes
	Agricultural and domestic residues
	Municipal solid wastes
	Organic industrial residues
	Sewage sludge
Reaction conditions	Temperature: 575–625 K
	Pressure: 12–18 MPa
	Resistance time: 5–20 min
	Medium: Liquid water
Main chemical reactions	Depolymerization
	Decarboxylation
	Dehydration
	Oxygen removed as CO_2 and H_2O
	Hydrodeoxygenation
	Hydrogenation
Products (%wt. on feedstock)	Biocrude: 45
	Water-soluble organics: 10
	Gas (> 90% CO_2): 25
	Process water: 20
Thermal efficiency	70–90%

2.5 Production of Fuels and Chemicals from Biomass

cellulose in water was obtained yielding hydrolysis products (cellobiose, glucose, etc., 44%) and decomposition products of glucose (erythrose, 1,6-anhydroglucose, 5-hydroxymethylfurfural, 3%). Furthermore, it has been shown that cellobiose decomposes via hydrolysis into glucose and via pyrolysis into glycosylerythrose and glycosylglycolaldehyde, which are further hydrolyzed into glucose, erythrose, and glycolaldehyde. Hydrolysis refers to the splitting up of organic particles into smaller organic fragments in water. Hydrothermal decomposition also acts on large organic molecules, reducing them into smaller fragments, some of which dissolve in water.

In the HTU process, biomass chips are pressurized and digested at 575 to 700 K with recycled water from the process. Subsequently the digested mass is pressurized to 12 to 18 MPa and reacted in liquid water at 575 to 700 K for 5 to 15 min. Under these conditions decarboxylation and depolymerization take place and a biocrude is formed that separates from the water phase. Part of the process water is recycled. Obviously, the process is very simple with high efficiency.

Hydrothermal reaction involves applying heat under pressure to achieve a reaction in an aqueous medium. The treatment of organic wastes by supercritical water (SCW) reaction occurs in a homogeneous phase where interface mass transfer limitations are avoided and reaction efficiencies of 99.9% can be achieved at residence times of less than 1 min. Because of the distinctive characteristics of water described above, hydrothermal reactions are an effective method for the treatment of organic wastes. The reactions can be performed under sub- or supercritical conditions. They can also be classified into two broad categories: (a) oxidative, i.e., involving the use of oxidants, and (b) nonoxidative, i.e., excluding the use of oxidants.

Pyrolysis oil (bio-oil) from wood is typically a liquid, almost black to dark red brown. The density of the liquid is about 1,200 kg/m^3, which is higher than that of fuel oil and significantly higher than that of the original biomass. Bio-oils typically have water contents of 14 to 33 wt%, which cannot be removed by conventional methods like distillation. Phase separation may occur above certain water contents. The higher heating value (HHV) is below 27 MJ/kg (compared to 43 to 46 MJ/kg for conventional fuel oils).

Pyrolysis is the thermal decomposition of organic matter occurring in the absence of oxygen or when significantly less oxygen is present than is required for complete combustion. Pyrolysis is the basic thermochemical process for converting biomass into a more useful fuel. Biomass is heated in the absence of oxygen, or partially combusted in a limited oxygen supply, to produce a hydrocarbon-rich gas mixture, an oil-like liquid, and a carbon-rich solid residue. The products of pyrolysis can be gaseous, liquid, or solid. Flash pyrolysis describes the rapid, moderate-temperature (675 to 875 K) pyrolysis that produces liquids. Biomass is heated at rates of 100 to 10,000 K/s, and the vapor residence time is normally less than 2 s. The oil products are maximized at the expense of char and gas. Pyrolysis processes typically use dry biomass sources.

Pyrolysis dates back to ancient Egyptian times, where tar for caulking boats and an embalming agent were made from pyrolysis. In the 1980s, researchers found that the pyrolysis liquid yield could be increased using fast pyrolysis, where a biomass feedstock is heated at a rapid rate and the vapors produced are also condensed rapidly (Mohan *et al.* 2006). Pyrolysis has been used since the dawn of civilization. If some means is applied to collect the off-gasses (smoke), the process is called wood distillation. The ancient Egyptians practiced wood distillation by collecting tars and pyroligneous acid for use in their embalming industry. Pyrolysis of wood to produce charcoal was a major industry in the 1800s, supplying the fuel for the industrial revolution, until it was replaced by coal. In the late 19th century and early 20th century wood distillation was still profitable for producing soluble tar, pitch, creosote oil, chemicals, and the noncondensable gases often used to heat boilers at the facility. The wood distillation industry declined in the 1930s due to the advent of the petrochemical industry and its lower-priced products.

Pyrolysis is a process similar to gasification except that it is generally optimized for the production of fuel liquids (pyrolysis oils) that can be used straight or refined for higher-quality uses such as engine fuels, chemicals, adhesives, and other products. Pyrolysis and the combustion of pyrolysis-derived fuel liquids and gases also produce the same categories of end products as the direct combustion of solids. Like gasification, their pollution control and conversion efficiencies may be improved.

Pyrolysis and direct liquefaction processes are sometimes confused with each other, and a simplified comparison of the two follows. Both are thermochemical processes in which feedstock organic compounds are converted into liquid products. In the case of liquefaction, feedstock macromolecule compounds are decomposed into fragments of light molecules in the presence of a suitable catalyst. At the same time, these fragments, which are unstable and reactive, repolymerize into oily compounds having appropriate molecular weights (Demirbas 2006b). With pyrolysis, on the other hand, a catalyst is usually unnecessary, and the light-decomposed fragments are converted into oily compounds through homogeneous reactions in the gas phase. The differences in operating conditions for liquefaction and pyrolysis are shown in Table 2.5.

Table 2.5 Comparison of liquefaction and pyrolysis

Process	Temperature (K)	Pressure (MPa)	Drying
Liquefaction	525–675	5–20	Unnecessary
Pyrolysis	675–975	0.1–0.5	Necessary

Catalytic cracking is a thermochemical process that employs catalysts using hydrogen-driven reducing reactions to accelerate the breakdown of high molecular weight compounds (e.g., plastics) into smaller products for the purposes of improving selectivity and imparting certain desirable characteristics to the final product, such as volatility and flashpoint of liquid fuels. This cracking process is often employed in oil refinery operations to produce lower molecular weight hydrocarbon fuels from waste feedstocks. These include gasoline from heavier oils, distillation residuals, and waste plastic.

In wood derived pyrolysis oil, specific oxygenated compounds are present in relatively large amounts. A current comprehensive review focuses on the recent de-

2.5 Production of Fuels and Chemicals from Biomass

velopments in the wood/biomass pyrolysis and reports the characteristics of the resulting bio-oils, which are the main products of fast wood pyrolysis (Mohan *et al.* 2006). Sufficient hydrogen added to the synthesis gas to convert all of the biomass carbon into methanol carbon would more than double the methanol produced from the same biomass base.

Rapid heating and rapid quenching produced the intermediate pyrolysis liquid products, which condense before further reactions break down higher-molecular-weight species into gaseous products. High reaction rates minimize char formation. Under some conditions, no char is formed. At higher fast pyrolysis temperatures, the major product is gas. Many researchers have attempted to exploit the complex degradation mechanisms by conducting pyrolysis in unusual environments (Mohan *et al.* 2006). The main pyrolysis variants are listed in Table 2.6.

Table 2.6 Pyrolysis methods and their variants

Method	Residence time	Temperature (K)	Heating rate	Products
Carbonation	Days	675	Very low	Charcoal
Conventional	5–30 min	875	Low	Oil, gas, char
Fast	0.5–5 s	925	Very high	Bio-oil
Flash-liquid[a]	< 1 s	< 925	High	Bio-oil
Flash-gas[b]	< 1 s	< 925	High	Chemicals, gas
Hydropyrolysis[c]	< 10 s	< 775	High	Bio-oil
Methanopyrolysis[d]	< 10 s	> 975	High	Chemicals
Ultra pyrolysis[e]	< 0.5 s	1,275	Very high	Chemicals, gas
Vacuum pyrolysis	2–30 s	675	Medium	Bio-oil

[a] Flash-liquid: liquid obtained from flash pyrolysis accomplished in a time of < 1 s.
[b] Flash-gas: gaseous material obtained from flash pyrolysis within a time of < 1 s.
[c] Hydropyrolysis: pyrolysis with water.
[d] Methanopyrolysis: pyrolysis with methanol.
[e] Ultra pyrolysis: pyrolysis with very high degradation rate.

Main feedstock preparation operations for biomass pyrolysis processes are cleaning, washing, handling, grinding, storing, transporting, and drying. Process variables of biomass pyrolysis are heat flux, temperature, partial pressure, feedstock, particle size, heating rates, residence time, heat and mass transfer, and mineral matter content. Process variables are important. For example, hot-water washing of biomass can improve oil quality and stability for high ash feedstocks. Heating rates, residence time, and heat and mass transfer are affected by particle size.

Activated carbons are carbonaceous materials with a highly developed internal surface area and porosity. Activated carbon is widely used as an effective adsorbent in many applications such as air separation and purification, vehicle exhaust emission control, solvent recovery, and catalyst support because of its high specific pore surface area, adequate pore size distribution, and relatively high mechanical strength. The large surface area results in high capacity for adsorbing chemicals from gases and liquids.

The starting materials used in commercial production of activated carbons are those with high carbon contents such as wood, lignite, peat, and coal of different ranks, or low-cost and abundantly available agricultural byproducts. Activated carbons can be manufactured from virtually any carbonaceous precursor, but the most commonly used materials are wood, coal, and coconut shell. The development of activated carbons from agricultural carbonaceous wastes will be advantageous for environmental problems. In water contamination, wastewater contains many traces of organic compounds, which are a serious environmental problem. In the development of activated carbons, agricultural carbonaceous wastes are used, as this will eliminate the problem of waste disposal while at the same time allowing societies to derive great economic benefits from such commercialized products.

Activated carbons are used in the following applications:

1. They can be used as adsorbents for the removal or reduction of gaseous pollutants from the exhaust gases of industrial sources.
2. They can be used as adsorbents for the removal of volatile organic compounds, ozone from air, mercury and dioxin emissions from incinerator flue gas, and hydrogen sulfide emissions from sewage treatment facilities.
3. They can be used to remove chlorine and organic chemicals (such as phenols, polychlorinated biphenyls, trihalomethanes, pesticides, and halogenated hydrocarbons), heavy metals, and organic contaminants from water.
4. They can be used to extract some harmful elements of cigarette smoke by incorporation in the filter tips of cigarettes.

Lignin gives higher yields of charcoal and tar from wood, although lignin has a threefold higher methoxyl content than wood. Phenolics are derived from lignin by cracking the phenyl-propane units of the macromolecule lattice. The formation of char from lignin under mild reaction conditions is a result of the breaking of the relatively weak bonds, like the alkyl–aryl ether bonds, and the consequent formation of more resistant condensed structures, as has already been noted.

Cellulose is a remarkable pure organic polymer consisting solely of units of anhydroglucose held together in a giant straight chain molecule. Cellulose must be hydrolyzed into glucose before fermentation into ethanol. Conversion efficiencies of cellulose into glucose may depend on the extent of chemical and mechanical pretreatments to structurally and chemically alter pulp and paper mill wastes. The method of pulping, the type of wood, and the use of recycled pulp and paper products could also influence the accessibility of cellulose to cellulase enzymes. Hemicelluloses (arabinoglycuronoxylan and galactoglucomammans) are related to plant gums in composition and occur in much shorter molecule chains than cellulose. The hemicelluloses, which are present in deciduous woods chiefly as pentosans and in coniferous woods almost entirely as hexosanes, undergo thermal decomposition very readily. Hemicelluloses are derived mainly from chains of pentose sugars and act as the cement material holding together the cellulose micells and fiber. Cellulose is insoluble in most solvents and has a low accessibility to acid and enzymatic hydrolysis. Hemicelluloses are largely soluble in alkali and, as such, are more easily hydrolyzed.

2.5 Production of Fuels and Chemicals from Biomass

Hydrolysis (saccharification) breaks down the hydrogen bonds in the hemicellulose and cellulose fractions into their sugar components: pentoses and hexoses. These sugars can then be fermented into bioethanol.

Cellulose [hexosan, $(C_6H_{10}O_5)_n$] hydrolysis produces glucose (a hexose, $C_6H_{12}O_6$). The hydrolysis of cellulose is catalyzed by mineral acids and enzymes.

Hexosan to hexose:

$$(C_6H_{10}O_5)_n + n\,H_2O \rightarrow n\,C_6H_{12}O_6 \qquad (2.1)$$
$$1\,g \qquad 0.111\,g \qquad 1.111\,g$$

Hemicellulose hydrolysis produces both hexose and pentose sugars – mannose, galactose, xylose, and arabinose – that are not all fermented with existing strains. The hemicelluloses fraction typically produces a mixture of sugars including xylose, arabinose, galactose, and mannose. Both of these are pentosans: xylose and arabinose, and hexosans: galactose and mannose. The hydrolysis of hemicelluloses is catalyzed by mineral acids and enzymes.

Pentosan to pentose:

$$(C_5H_8O_4)_n + n\,H_2O \rightarrow n\,C_5H_{10}O_5 \qquad (2.2)$$
$$1\,g \qquad 0.136\,g \qquad 1.136\,g$$

The most commonly applied methods can be classified into two groups: chemical hydrolysis (dilute and concentrated acid hydrolysis) and enzymatic hydrolysis. In addition, there are other hydrolysis methods in which no chemicals or enzymes are applied. For instance, lignocellulose may be hydrolyzed by gamma-ray or electron-beam irradiation, or by microwave irradiation. However, those processes are commercially unimportant.

Both enzymatic and chemical hydrolyses require a pretreatment to increase the susceptibility of cellulosic materials. In chemical hydrolysis, the pretreatment and the hydrolysis may be carried out in a single step. There are two basic types of acid hydrolysis processes commonly used: dilute acid and concentrated acid, each with variations.

The dilute acid process is conducted under high temperature and pressure and has a reaction time in the range of seconds or minutes, which facilitates continuous processing. As an example, using a dilute acid process with 1% sulfuric acid in a continuous-flow reactor at a residence time of 0.22 min and a temperature of 510 K with pure cellulose provided a yield of over 50% sugars. In this case, 1,000 kg of dry wood would yield about 164 kg of pure ethanol. The combination of acid and high temperature and pressure dictates special reactor materials, which can make the reactor expensive. The first reaction converts the cellulosic materials into sugar and the second reaction converts the sugars into other chemicals. Unfortunately, the conditions that cause the first reaction to occur are also the right conditions for the second to occur.

The biggest advantage of dilute acid processes is their fast rate of reaction, which facilitates continuous processing. Since 5-carbon sugars degrade more rapidly than

6-carbon sugars, one way to decrease sugar degradation is to have a two-stage process. The first stage is conducted under mild process conditions to recover the 5-carbon sugars, while the second stage is conducted under harsher conditions to recover the 6-carbon sugars.

Hydrolysis of cellulosic materials by concentrated sulfuric or hydrochloric acid is a relatively old process. The concentrated acid process uses relatively mild temperatures, and the only pressures involved are those created by pumping materials from vessel to vessel. Reaction times are typically much longer than for dilute acid. This method generally uses concentrated sulfuric acid followed by a dilution with water to dissolve and hydrolyze or convert the substrate into sugar. This process provides a complete and rapid conversion of cellulose to glucose and hemicelluloses into 5-carbon sugars with little degradation. The critical factors needed to make this process economically viable are the optimization of sugar recovery and cost-effective recovery of the acid for recycling. The solid residue from the first stage is dewatered and soaked in a 30 to 40% concentration of sulfuric acid for 1 to 4 h as a precellulose hydrolysis step. The solution is again dewatered and dried, increasing the acid concentration to about 70%. After reacting in another vessel for 1 to 4 h at low temperatures, the contents are separated to recover the sugar and acid. The sugar/acid solution from the second stage is recycled to the first stage to provide the acid for the first stage of hydrolysis.

The primary advantage of the concentrated acid process is the potential for high sugar recovery efficiency. The acid and sugar are separated via ion exchange and then the acid is reconcentrated via multiple effect evaporators. The low temperatures and pressures employed allow the use of relatively low-cost materials such as fiberglass tanks and piping. The low temperatures and pressures also minimize the degradation of sugars. Unfortunately, it is a relatively slow process and cost-effective acid recovery systems have been difficult to develop. Without acid recovery, large quantities of lime must be used to neutralize the acid in the sugar solution. This neutralization leads to the formation of large quantities of calcium sulfate, which requires disposal and creates additional expense.

Another basic method of hydrolysis is enzymatic hydrolysis. Enzymes are naturally occurring plant proteins that cause certain chemical reactions to occur. There are two technological developments important for this process: enzymatic and direct microbial conversion methods.

The chemical pretreatment of the cellulosic biomass is necessary before enzymatic hydrolysis. The first application of enzymatic hydrolysis was used in separate hydrolysis and fermentation steps. Enzymatic hydrolysis is accomplished by cellulolytic enzymes. Different kinds of "cellulases" may be used to cleave the cellulose and hemicelluloses. A mixture of endoglucanases, exoglucanases, β-glucosidases, and cellobiohydrolases is commonly used. The endoglucanases randomly attack cellulose chains to produce polysaccharides of shorter length, whereas exoglucanases attach to the nonreducing ends of these shorter chains and remove cellobiose moieties. β-glucosidases hydrolyze cellobiose and other oligosaccharides to glucose.

For enzymes to work efficiently, they must obtain access to the molecules to be hydrolyzed. This requires some kind of pretreatment process to remove hemicel-

2.5 Production of Fuels and Chemicals from Biomass

luloses and break down the crystalline structure of the cellulose or removal of the lignin to expose hemicelluloses and cellulose molecules.

The main wood valorization technologies include pulp and paper making, bio-oil by pyrolysis, synthesis gas by gasification, sugar by hydrolysis, ethanol by sugar fermentation, and adhesives by alkali liquefaction and polymerization.

Adhesion is a state in which two surfaces are held together by interfacial forces, which may be valence forces, interlocking action, or both. The adhesives used in commercial wood composite products are usually synthetic polymer resins, based on the condensation reaction of formaldehyde with phenol, urea, resorcinol, or melamine. More than 70% of this urea–formaldehyde resin is used by the forest products industry for a variety of purposes.

The wood adhesive market is very large and problems due to volatile organic compounds and toxic chemicals in many adhesives and their production are significant. In addition, most of the adhesives are derived from depleting petrochemical resources. An environmentally friendly wood adhesive based on renewable resources and produced by microbial fermentation has been explored.

A method is described for making adhesive from biomass. The liquefaction oil is prepared from lignin-bearing plant material and a phenolic fraction is extracted therefrom. The phenolic fraction is reacted with formaldehyde to yield a phenol–formaldehyde resin. At present, the production of wood composites mainly relies on petrochemical-based and formaldehyde-based adhesives such as PF resins and urea–formaldehyde (UF) resins. Phenol–formaldehyde adhesives are used to manufacture plywood, flakeboard, and fiberglass insulation. Phenolic resins occur only in dark, opaque colors and can therefore be used only to manufacture dark-colored products.

Resistance to chemical attack is generally improved by resin impregnation, which protects the underlying wood and reduces movement of liquid into the wood. Resistance to acids can be obtained by impregnating with phenolic resin and to alkalies by impregnating with furfural resin.

The adhesion properties of different natural fillers without the addition of coupling agents by considering the different filler morphology were investigated. The adhesion behavior was determined in a qualitative way from microscopic observation and it was shown that its influence on mechanical properties decreases, since in this type of material tensile strength usually falls.

The reaction of formaldehyde with lignin model compounds in acidic medium is shown to give fast cross-linking of alkali-substituted phenolic and etherified phenolic lignin model compounds at positions meta to the aromatic hydroxyl groups. This reaction differs from the reaction of formaldehyde with phenolic lignin model compounds in alkaline conditions, where the reaction with formaldehyde always occurs at positions ortho/para to the aromatic hydroxyl group.

Lignin degradation products and their sodium salts can be converted into very weak organic acids by treating mineral acids. Most phenols have a Ka in the neighborhood of 10^{-10} and are thus considerably weaker than the carboxylic acids (Ka about 10^{-5}). Most phenols are weaker than carbonic acid ($CO_2 + H_2O$), and, hence, unlike carboxylic acids, do not dissolve in aqueous bicarbonate solutions. Indeed,

46   2 Green Energy Facilities

phenols are conveniently liberated from their salts by the action of carbonic acid
(Ka about 10^{-7}).

Organosolv lignin-phenol-formaldehyde (LPF) resins were produced in a two-step preparation with different additions of lignin. The method selected for the manufacture of lignin resins dealt with modification of the lignin by the methylolation route. Organosolv lignin-based resins showed comparatively good strength and stiffness. The tensile strength properties of test samples made from organosolv lignin resins were equal to or better than those of test samples made from PF resin only.

References

Alanne, K., Sari, A. 2006. Distributed energy generation and sustainable development. Renew Sust Energy Rev 10:539–558.
Arkesteijn, K., Oerlemans, L. 2005. The early adoption of green power by Dutch households: an empirical exploration of factors influencing the early adoption of green electricity for domestic purposes. Energy Policy 33:183–196.
Boerrigter, H., den Uil, H. 2002. Green diesel from biomass via FTS: new insights in gas cleaning and process design. Pyrolysis and Gasification of Biomass and Waste, Expert Meeting, Strasbourg, France, 30 Sep.–1 Oct.
Demirbas, A. 2006b. Biomass gasification for power generation in Turkey. Energy Sources Part A 28:433–445.
Demirbas, A. 2006c. Biofuel based cogenerative energy conversion systems. Energy Sources Part A 28:1509–1518.
Demirbas, A. H., Demirbas, I. 2007. Importance of rural bioenergy for developing countries. Energy Convers Manage 48:2386–2398.
Elliott, D. 1999. Prospects for renewable energy and green energy markets in the UK. Renew Energy 16:1268–1271.
Feng, W., van der Kooi, H. J., Arons, J. D. S. 2004. Biomass conversions in subcritical and supercritical water: driving force, phase equilibria, and thermodynamic analysis. Chem Eng Proc 43:1459–1467.
Fridleifsson, I. B. 2003. Status of geothermal energy amongst the world's energy sources. Geothermics 32:379–388.
Goudriaan, F., Peferoen, D. 1990. Liquid fuels from biomass via a hydrothermal process. Chem Eng Sci 45:2729–2734.
Haberl, H., Geissler, S. 2000. Cascade utilization of biomass: strategies for a more efficient use of a scarce resource. Ecol Eng 16:S111–S121.
Hall, D. O., Rosillo-Calle, F., Williams, R. H., Woods, J. 1993. Biomass energy supply and prospects. In: Johansson, T. B., Kelly, H., Reddy, A. K. N., Williams, R. H. (eds.) Renewable Energy: Sources for Fuel and Electricity. Island Press, Washington, D.C., pp. 593–651.
Hoogwijk, M., Faaij, A., van den Broek, R., Berndes, G., Gielen, D., Turkenburg, W. 2003. Exploration of the ranges of the global potential of biomass for energy. Biomass Bioenergy 25:119–133.
Itoh, S., Suzuki, A., Nakamura, T., Yokoyama, S. 1994. Production of heavy oil from sewage sludge by direct thermochemical liquefaction. In: Proceedings of the IDA and WRPC World Conference on Desalination and Water Treatment, vol. 98, pp. 127–133.
Inoue, S., Sawayma, S., Dote, Y., Ogi, T. 1997. Behavior of nitrogen during liquefaction of dewatered sewage sludge. Biomass Bioenergy 12:473–475.
Jin, Y., Datye, A. K. 2000. Phase transformations in iron Fischer–Tropsch catalysts during temperature-programmed reduction. J Catal 196:8–17.

References

Karki, S., Mann, M. D., Salehfar, H. 2008. Environmental implications of renewable distributed generation technologies in rural electrification. Energy Sources Part B 3:186–195.

Kranich, W. L. 1984. Conversion of sewage sludge to oil by hydroliquefaction. EPA-600/2 84-010. Report for the U.S. Environmental Protection Agency. Cincinnati, OH. U.S. EPA.

Larson, E. D. 1993. Technology for fuels and electricity from biomass. Annu Rev Energy Environ 18:567–630.

Mohan, D., Pittman Jr., C. U., Steele, P. H. 2006. Pyrolysis of wood/biomass for bio-oil: a critical review. Energy Fuels 2006;20:848–889.

Murphya, H., Niitsuma, H. 1999. Strategies for compensating for higher costs of geothermal electricity with environmental benefits. Geothermics 28:693–711.

Pimentel, D., Moran, M. A., Fast, S., Weber, G., Bukantis, R., Balliett, L., Boveng, P., Cleveland, C., Hindman, S., Young, M. 1981. Biomass energy from crop and forest residues. Science 212:1110–1115.

Suzuki, A., Yokoyama, S., Murakami, M., Ogi, T., Koguchi, K. 1986. New treatment of sewage sludge by direct thermochemical liquefaction. Chem Lett CMLTAG.I 9:1425–1428.

Tijmensen, M. J. A., Faaij, A. P. C., Hamelinck, C. N., van Hardeveld, M. R. M. 2002. Exploration of the possibilities for production of Fischer Tropsch liquids and power via biomass gasification. Biomass Bioenergy 23:129–152.

UNDP 2000. World Energy Assessment. Energy and the challenge of sustainability. United Nations Development Programme, New York.

Zhang, R., Zhang, Z. 1999. Biogasification of rice straw with an anaerobic-phased solids digester system. Biores Technol 68:235–245.

Chapter 3
Biofuels

3.1 Introduction

Today's energy system is unsustainable because of equity issues as well as environmental, economic, and geopolitical concerns that have implications far into the future. Bioenergy is one of the most important components to mitigate greenhouse gas emissions and substitute for fossil fuels (Goldemberg 2000; Dincer 2008). Renewable energy is one of the most efficient ways to achieve sustainable development.

Plants use photosynthesis to convert solar energy into chemical energy. It is stored in the form of oils, carbohydrates, proteins, etc. This plant energy can be converted to biofuels. Hence biofuels are primarily a form of solar energy. For biofuels to succeed at replacing large quantities of petroleum fuel, the feedstock availability needs to be as high as possible.

In the context of climatic changes and of soaring prices for a barrel of petroleum, biofuels are now being presented as a renewable energy alternative. Presently, research is being done on microscopic algae, or microalgae, which are particularly rich in oils and whose yield per hectare is considerably higher than that of sunflower or rapeseed.

In recent years, recovery of liquid transportation biofuels from biorenewable feedstocks has became a promising method. The biggest difference between biorenewable and petroleum feedstocks is oxygen content. Biorenewables have oxygen levels of 10 to 44% while petroleum has essentially none, making the chemical properties of biorenewables very different from those of petroleum (Demirbas 2008; Balat 2009). For example, biorenewable products are often more polar and some easily entrain water and can therefore be acidic.

There are two global transportation fuels – gasoline and diesel fuel. The main transportation fuels that can be obtained from biomass using different processes are sugar ethanol, cellulosic ethanol, grain ethanol, biodiesel, pyrolysis liquids, green diesel, green gasoline, butanol, methanol, syngas liquids, biohydrogen, algae diesel, algae jet fuel, and hydrocarbons. Renewable liquid biofuels for transportation have recently attracted considerable attention in different countries around the world be-

A. Demirbas, M. Fatih Demirbas, *Algae Energy*
DOI 10.1007/978-1-84996-050-2, © Springer 2010

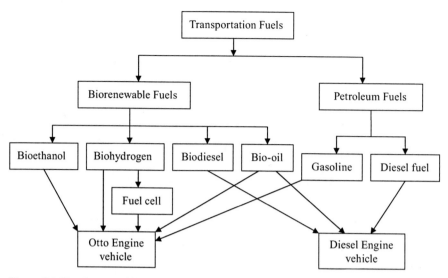

Figure 3.1 Petroleum- and biorenewable-based transportation fuels

cause of their renewability, sustainability, widespread availability, and biodegradability, as well as the benefits they bring with respect to regional development, rural manufacturing jobs, and reduction in greenhouse gas emissions (Demirbas 2008). Petroleum- and biorenewable-based transportation fuels are given in Figure 3.1.

The policy drivers for biorenewable liquid biofuels have been rural development and economic opportunities for developing countries (Keskin 2009). The European Union ranks third in biofuel production worldwide, behind Brazil and the USA. In Europe, Germany is the largest and France the second largest producer of biofuels.

The term modern biomass is generally used to describe traditional biomass use through efficient and clean combustion technologies and sustained supply of biomass resources, environmentally sound and competitive fuels, heat, and electricity using modern conversion technologies. As an energy source, biomass has two striking characteristics. First, biomass is the only renewable organic resource as well as one of the most abundant resources. Second, biomass fixes carbon dioxide in the atmosphere by photosynthesis. Direct combustion and cofiring with coal for electricity production from biomass holds great promise for the near future. Biomass thermochemical conversion technologies such as pyrolysis and gasification are certainly not the most important options at present; combustion is responsible for over 97% of the world's bioenergy production. Ethanol and fatty acid (m)ethylester (biodiesel) as well as diesel produced from biomass by Fischer–Tropsch synthesis (FTS) are modern biomass-based transportation fuels. Liquid transportation fuels can be economically produced by Biomass Integrated Gasification Fischer–Tropsch (BIG-FT) processes. Modern biomass produced in a sustainable way excludes traditional uses of biomass as fuel wood and includes electricity generation and heat production, as well as transportation fuels, from agricultural and forest residues and

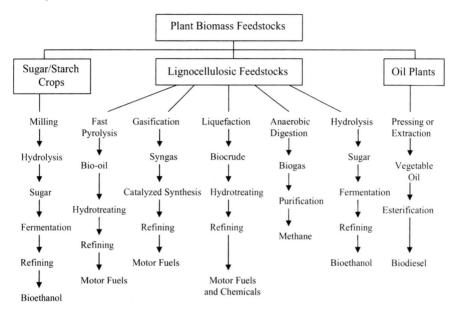

Figure 3.2 Overview of conversion processes of plant biomass feedstocks into biofuels

solid waste. On the other hand, "traditional biomass" is produced in an unsustainable way and is used as a noncommercial source – usually with very low efficiencies for cooking in many countries.

A biorefinery is an integrated plant producing multiple value-added products from a range of renewable feedstocks. This innovative approach responds to changing markets for traditional forest products as well as new products such as energy, chemicals, and materials. The range of feedstocks, processes, and potential products is large; each combination of feedstock, process, and product is characterized by its own unique combination of technical and economic opportunities, emerging technologies, and barriers. Figure 3.2 shows an overview of conversion routes of plant biomass feedstocks to biofuels.

3.2 Importance of Biofuels

Liquid biofuels will be important in the future because they replace petroleum fuels. The biggest difference between biofuels and petroleum feedstocks is oxygen content. Biofuels are nonpolluting, locally available, accessible, sustainable, and reliable fuel obtained from renewable sources. Biofuels can be classified based on their production technologies: first-generation biofuels (FGBs), second-generation biofuels (SGBs), third-generation biofuels (TGBs), and fourth-generation biofuels.

The FGBs refer to biofuels made from sugar, starch, vegetable oils, or animal fats using conventional technology. The basic feedstocks for the production of FGBs are often seeds or grains such as wheat, which yields starch that is fermented into bioethanol, or sunflower seeds, which are pressed to yield vegetable oil that can be used in biodiesel.

SGBs and TGBs are also called advanced biofuels. SGBs are made from nonfood crops, wheat straw, corn, wood, and energy crops using advanced technology. Algae fuel, also called algal oil or a TGB, is a biofuel from algae. Algae are low-input/high-yield (30 times more energy per acre than land) feedstocks to produce biofuels using more advanced technology. On the other hand, an emerging fourth-generation fuel is based on the conversion of vegetable oil and biodiesel into biogasoline using the most advanced technology.

The SGBs include renewable and green diesels. The former involves a technology that incorporates vegetable oils in the crude-oil-derived diesel production process to produce a renewable carbon-based diesel with no oxygen content and a very high cetane number, while the latter entails the production of middle distillate by means of Fischer–Tropsch (FT) catalysts, using synthesis gas produced by the gasification of biomass. FT-like catalysts (synthol process) can also produce ethanol and mixed alcohols.

There are some barriers to the development of biofuel production. They are technological, economic, supply, storage, safety, and policy barriers. Reducing these barriers is one of the driving factors in government involvement in biofuel and biofuel research and development. Production costs are uncertain and vary with the feedstock available. The production of biofuels from lignocellulosic feedstocks can be achieved through two very different processing routes: biochemical and thermochemical. There is no clear candidate for "best technology pathway" between the competing biochemical and thermochemical routes. Technical barriers for enzymatic hydrolysis include low specific activity of current commercial enzymes, high cost of enzyme production, and lack of understanding of enzyme biochemistry and mechanistic fundamentals.

The major nontechnical barriers are restrictions or prior claims on of land use (food, energy, amenity use, housing, commerce, industry, leisure or designations as areas of natural beauty, special scientific interest, etc.), as well as the environmental and ecological effects of large areas of monoculture. For example, vegetable oils are a renewable and potentially inexhaustible source of energy with energy content close to that of diesel fuel. On the other hand, extensive use of vegetable oils may cause other significant problems such as starvation in developing countries. The vegetable oil fuels were not acceptable because they were more expensive than petroleum fuels.

There are few technical barriers to building biomass-fired facilities at any scale, from domestic to around 50 MW, above which considerations of the availability and cost of providing fuel become significant. In general, however, the capacity and generating efficiency of biomass plants are considerably less than those of modern natural-gas-fired turbine systems. The main nontechnical limitations to investment in larger systems are economic, or in some countries reflect planning conditions and public opinion, where a clear distinction may not be made between modern effective biomass energy plant and older polluting incinerator designs.

3.2 Importance of Biofuels

The most important biorenewable liquid fuels are bioethanol and biodiesel. Bioethanol is a petrol additive/substitute. Biodiesel is a diesel alternative. Biorenewable fuels are safely and easily biodegradable and so are particularly attractive from an environmental perspective. Biodiesel, a biofuel that can directly replace petroleum-derived diesel without engine modifications, has gained a lot of attention due to its environmental and technological advantages.

Production of motor fuel alternatives from biomass materials is an important application area of biotechnological methods. Table 3.1 shows the potential and available motor fuels. Biorenewable sourced motor fuel alternatives are:

1. Gasoline-alcohol mixtures
2. Alcohol substituting for gasoline
3. Gasoline-vegetable oil mixtures
4. Diesel fuel-vegetable oil mixtures
5. Vegetable oil substituting for diesel fuel.

Table 3.1 Potential and available motor fuels

Fuel type	Available motor fuel
Traditional fuels	Diesel and gasoline
Oxygenated fuels	Ethanol 10% (E10), methanol, methyl tertiary butyl ether (MTBE), ethyl tertiary butyl ether (ETBE), tertiary butyl alcohol (TBA), and tertiary amyl methyl ether (TAME)
Alternative fuels	Liquefied petroleum gases (LPG), ethanol, 85% (E85), ethanol, 95% (E95), methanol, 85% (M85), methanol, neat (M100), compressed natural gas (CNG), liquefied natural gas (LNG), biodiesel (BD), hydrogen, and electricity

In gasoline-alcohol mixtures ethanol and methanol are generally used, and in gasoline engine mixtures containing 20% or less alcohol by volume can be used without altering the construction of the engine. Because of the hygroscopic properties of ethanol and methanol, gasoline-alcohol mixtures are in fact ternary mixtures composed of gasoline-alcohol and water. In the evaluation of such mixtures as motor fuel, there is the phase separation problem, which depends on several factors. It is evident in the literature that numerous attempts have been made to overcome this problem (Mislavskaya *et al.* 1982; Osten and Sell 1983).

In gasoline-methanol mixtures containing 0.1% water i-propanol is added to the environment (medium) in order to decrease the phase separation temperature, and fuels containing different ratios of gasoline-methanol-i-propanol and water are composed that have proven to be stable in certain climatic conditions. An increase in the aromatic character of the gasoline, a decrease in the water content of the mixture, and an increase in the amount of the additive used results in a decrease in the phase separation temperature of the mixture. In gasoline-ethanol mixtures the additive used is also i-propanol. In gasoline-alcohol mixtures various additives like i-propanol, n-butanol, i-butanol, and i-amylalcohol are used.

3.3 Bioethanol

Ethanol (CH_5OH or EtOH) is a liquid biofuel that can be produced from several different biomass feedstocks and conversion technologies. Bioethanol is an attractive alternative fuel because it is a renewable bio-based resource and is oxygenated, thereby providing the potential to reduce particulate emissions in compression-ignition engines.

Ethanol can be made synthetically from petroleum or by microbial conversion of biomass materials through fermentation. In 1995, about 93% of the ethanol in the world was produced by the fermentation method and about 7% by the synthetic method. The fermentation method generally uses three steps: (1) the formation of a solution of fermentable sugars, (2) the fermentation of these sugars to ethanol, and (3) the separation and purification of the ethanol, usually by distillation.

Ethanol has a higher octane number, broader flammability limits, higher flame speeds, and higher heats of vaporization than gasoline. These properties allow for a higher compression ratio, shorter burn time and leaner burn engine, which lead to theoretical efficiency advantages over gasoline in an internal combustion engine. The disadvantages of ethanol include its lower energy density than gasoline (but about 35% higher than that of methanol), its corrosiveness, low flame luminosity, lower vapor pressure (making cold starts difficult), miscibility with water, and toxicity to ecosystems.

Because ethanol-based fuel contains oxygen (35% oxygen content), it can effectively reduce particulate matter emission in a diesel engine. Ethanol is appropriate for mixed fuel in gasoline engines because of its high octane number, and its low cetane number and high heat of vaporization impede self-ignition in diesel engines. The most popular blend for light-duty vehicles is known as E85, which contains 85% ethanol and 15% gasoline. In Brazil, bioethanol for fuel is derived from sugar cane and is used pure or blended with gasoline in a mixture called gasohol (24% bioethanol, 76% gasoline). In several states of the USA, a small amount of bioethanol (10% by volume) is added to gasoline; this is known as gasohol or E10. Blends having higher concentrations of bioethanol in gasoline are also used, e.g., in flexible-fuel vehicles (FFVs) that can operate on blends of up to 85% bioethanol – E85. Some countries have established biofuel programs on bioethanol-gasoline blends such as the USA (E10 and for flexible fuel vehicle [FFV] E85), Canada (E10 and for FFV E85), Sweden (E5 and for FFV E85), India (E5), Australia (E10), Thailand (E10), China (E10), Colombia (E10), Peru (E10), Paraguay (E7), and Brazil (E20, E25, and FFV any blend) (Kadiman 2005).

As biomass hydrolysis and sugar fermentation technologies approach commercial viability, advancements in product recovery technologies will be required. For cases in which fermentation products are more volatile than water, recovery by distillation is often the technology of choice. Distillation technologies that will allow the economical recovery of dilute volatile products from streams containing a variety of impurities have been developed and commercially demonstrated. A distillation system separates the bioethanol from water in the liquid mixture.

3.3 Bioethanol

The first step is to recover the bioethanol in a distillation or beer column, where most of the water remains with the solid parts. The product (37% bioethanol) is then concentrated in a rectifying column to a concentration just below the azeotrope (95%). The remaining bottoms product is fed to the stripping column to remove additional water, with the bioethanol distillate from stripping being recombined with the feed to the rectifier. The recovery of bioethanol in the distillation columns in the plant is fixed at 99.6% to reduce bioethanol losses.

Bioethanol can be used directly in cars designed to run on pure ethanol or blended with gasoline to make "gasohol." Anhydrous ethanol is required for blending with gasoline. No engine modification is typically needed to use the blend. Ethanol can be used as an octane-boosting, pollution-reducing additive in unleaded gasoline.

Figure 3.3 shows the world production by country of ethanol between 1980 and 2008 (RFA 2009). Between 1991 and 2001, world ethanol production rose from around 16 billion liters a year to 18.5 billion liters. From 2001 to 2007, production is expected to have tripled, to almost 60 billion liters a year. Brazil was the world's leading ethanol producer until 2005, when US production roughly equaled Brazil's. The USA became the world's leading ethanol producer in 2006. China holds a distant but important third place in world rankings, followed by India, France, Germany, and Spain. Ethanol production by country in 2007 and 2008 is given in Table 3.2.

The continued increases in the price of crude oil in 2005 and 2006 resulted in a reversal of the traditional relationship between the price of biomass energy and that of crude oil, something not seen since the 1930s. As a consequence of the high prices of traded crude oil, many countries advanced their biofuel goals, and, in the case of Brazil and the USA, large production gains occurred.

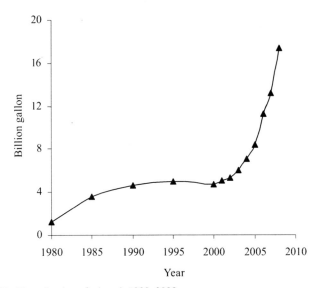

Figure 3.3 World production of ethanol, 1980–2008

Table 3.2 Ethanol production by country, 2007–2008 (millions of gallons, 1 gallon = 3.78 liters)

Country	2007	2008
USA	6,500	9,001
Brazil	5,020	6,473
European Union	571	734
China	487	502
Canada	212	238
Others	118	130
Thailand	80	90
Colombia	75	80
India	53	66
Australia	26	27
Turkey	16	18
Total	13,158	17,359

3.4 Biodiesel

Starting in the 1980s important progress has been made on evaluating some low-grade oils, oil production wastes, and residues as motor fuel (Pryor *et al.* 1983). But direct usage of vegetable oils causes a number of problems concerning the engine because of their high viscosity and the excessive carbonaceous deposits left in the cylinders and on the injector nozzles. Therefore, chemical conversion of vegetable oils was suggested. In order to lower the viscosities and flash points of vegetable oils, the transesterification method was applied, and it was reported that the alcoholysis products of soybean, sunflower, rapeseed, and used frying oils were proposed as diesel fuel alternatives.

The deregulation of domestic crude oil prices and the formation of OPEC have been largely responsible for high fuel prices. The farmer is highly dependent on diesel fuel for crop production. Alternative fuels such as vegetable oils could help ease the petroleum dependence of farmers. Recently, the demand for crude oil has decreased because of conservation practices but ultimately a liquid fuel resource problem exists (Pryor *et al.* 1983).

Vegetable oils with carbon chain lengths of between 16 and 22 carbon atoms are generally in the form of triacyl glycerides (TAG), which upon transesterification with methanol produce fatty acid methyl ester (FAME) as the precursor to biodiesel and glycerol as a byproduct. Vegetable oil (m)ethyl esters, commonly referred to as "biodiesel," are prominent candidates as alternative diesel fuels. The name biodiesel has been given to transesterified vegetable oil to describe its use as a diesel fuel. After FAME purification and testing for compliance with either EN 14214 or ASTM D6751 standards the product can be sold as biodiesel and used as blends – typically B5 (5% biodiesel) to B20, depending on the engine warranties.

Biodiesel is a mixture of methyl esters of long-chain fatty acids like lauric, palmitic, steric, oleic, etc. Typical examples are rapeseed oil, canola oil, soybean oil, sunflower oil, and palm oil and its derivatives from vegetable sources. Beef

3.4 Biodiesel 57

and sheep tallow and poultry oil from animal sources and cooking oil are also the sources of raw materials. The chemistry of conversion to biodiesel is essentially the same. Oil or fat reacts with methanol or ethanol in the presence of catalyst sodium hydroxide or potassium hydroxide to form biodiesel, (m)ethylesters, and glycerin. Biodiesel is technically competitive with or offers technical advantages over conventional petroleum diesel fuel. Biodiesel esters are characterized for their physical and fuel properties including density, viscosity, iodine value, acid value, cloud point, pure point, gross heat of combustion, and volatility. Biodiesel fuels produce slightly lower power and torque, and higher fuel consumption than No. 2 diesel fuel. Biodiesel is better than diesel fuel in terms of sulfur content, flash point, aromatic content, and biodegradability. Some technical properties of biodiesels are shown in Table 3.3.

Table 3.3 Some technical properties of biodiesels

Common name	Bio-diesel or biodiesel
Common chemical name	Fatty acid (m)ethyl ester
Chemical formula range	C_{14}–C_{24} methyl esters or C_{15-25} H_{28-48} O_2
Kinematic viscosity range (mm^2/s at 313 K)	3.3–5.2
Density range (kg/m^3 at 288 K)	860–894
Boiling point range (K)	> 475
Flash point range (K)	428–453
Distillation range (K)	470–600
Vapor pressure (mm Hg, at 295 K)	< 5
Solubility in water	Insoluble in water
Physical appearance	Light to dark yellow, clear liquid
Odor	Light musty/soapy odor
Biodegradability	More biodegradable than petroleum diesel
Reactivity	Stable, but avoid strong oxidizing agents

Biodiesel is a synthetic diesel-like fuel produced from vegetable oils, animal fats, or waste cooking oil. It can be used directly as fuel, which requires some engine modifications, or blended with petroleum diesel and used in diesel engines with few or no modifications. At present, biodiesel accounts for less than 0.2% of the diesel consumed for transport (UN 2006). Biodiesel has become more attractive recently because of its environmental benefits. The cost of biodiesel, however, is the main obstacle to commercialization of the product. With cooking oils used as raw material, the viability of a continuous transesterification process and recovery of high-quality glycerol as a biodiesel byproduct are the primary methods to be considered for lowering the cost of biodiesel (Zhang *et al.* 2003). Table 3.4 shows the biodiesel production (2007) and production capacity (2008) of EU countries (EBB 2009). Figure 3.4 shows the world production of biodiesel between 1980 and 2008 (Demirbas 2008).

Between 1991 and 2001, world biodiesel production grew steadily to approx. 1 billion liters. Most of this production was in OECD Europe and was based on virgin vegetable oils. Small plants using waste cooking oils started to be built in

Table 3.4 Biodiesel production (2007) and production capacity (2008) of EU countries (millions of tons)

Country	2007 (production)	2008 (production capacity)
Germany	2,890	5,302
France	872	1,980
Italy	363	1,566
Spain	168	1,267
UK	150	726
Belgium	166	665
The Netherlands	85	571
Greece	100	565
Austria	267	485
Poland	80	450
Portugal	175	406
Bulgaria	9	215
Sweden	63	212
Slovakia	46	206
Czech Republic	61	203
Hungary	7	186
Finland	39	170
Lithuania	26	147
Denmark	85	140
Estonia	0	135
Latvia	9	130
Romania	36	111
Slovenia	11	67
Ireland	3	80
Malta	1	8
Cyprus	1	6
Luxembourg	0	0
Total	5,713	16,000

other OECD countries by the late 1990s, but the industry outside Europe remained insignificant until around 2004. Since then, governments around the world have instituted various policies to encourage development of the industry, and new capacity in North America, Southeast Asia, and Brazil has begun to come onstream at a brisk rate. As a result, between 2001 and 2007, biodiesel production will have grown almost tenfold, to 9 billion L (Demirbas 2008).

The advantages of biodiesel as diesel fuel are its portability, ready availability, renewability, higher combustion efficiency, lower sulfur and aromatic content (Knothe et al. 2006; Demirbas 2008), higher cetane number, and higher biodegradability (Zhang et al. 2003). The main advantages of biodiesel given in the literature include its domestic origin, reducing the dependency on imported petroleum, biodegradability, high flash point, and inherent lubricity in the neat form (Knothe et al. 2005). The disadvantages of biodiesel are its higher viscosity, lower energy content, higher cloud point and pour point, higher nitrogen oxide (NO_x) emissions, lower engine

3.5 Bio-oil

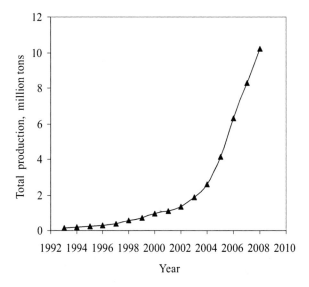

Figure 3.4 World production of biodiesel, 1980–2008

speed and power, injector coking, engine compatibility, high price, and higher engine wear.

The cost of feedstock is a major economic factor in the viability of biodiesel production. Using an estimated process cost (Note: all costs are given in US$), exclusive of feedstock cost, of $0.158/L ($0.60/gal) for biodiesel production, and estimating a feedstock cost of $0.539/L ($2.04/gal) for refined soy oil, an overall cost of $0.70/L ($2.64/gal) for the production of soy-based biodiesel was estimated (Haas *et al.* 2006). Biodiesel from animal fat is currently the cheapest option ($0.4 to $0.5/L), while traditional transesterification of vegetable oil is at present around $0.6 to $0.8/L (IEA 2007). Rough projections of the cost of biodiesel from vegetable oil and waste grease are, respectively, $0.54 to $0.62/L and $0.34 to $0.42/L. With pretax diesel priced at $0.18/L in the USA and $0.20 to 0.24/L in some European countries, biodiesel is thus currently not economically feasible, and more research and technological development will be needed (Bender 1999).

3.5 Bio-oil

The term bio-oil is used mainly to refer to liquid fuels from biorenewable feedstocks. Biomass is heated in the absence of oxygen, or partially combusted in a limited oxygen supply, to produce an oil-like liquid, a hydrocarbon-rich gas mixture and a carbon-rich solid residue. Pyrolysis dates back to at least ancient Egyptian times, when tar for caulking boats and certain embalming agents was made by pyrolysis. In the 1980s, researchers found that the pyrolysis liquid yield could be increased us-

60 3 Biofuels

ing fast pyrolysis where a biomass feedstock is heated at a rapid rate and the vapors produced are also condensed rapidly (Mohan *et al.* 2006).

In wood-derived pyrolysis oil, specific oxygenated compounds are present in relatively large amounts. A current comprehensive review focuses on the recent developments in wood/biomass pyrolysis and reports the characteristics of the resulting bio-oils, which are the main products of fast wood pyrolysis (Mohan *et al.* 2006). Sufficient hydrogen added to the synthesis gas to convert all of the biomass carbon into methanol carbon would more than double the methanol produced from the same biomass base.

The kinematic viscosity of pyrolysis oil varies from as low as 11 cSt to as high as 115 mm^2/s (measured at 313 K) depending on the nature of the feedstock, the temperature of the pyrolysis process, the thermal degradation degree and catalytic cracking, the water content of the pyrolysis oil, the amount of light ends that have collected, and the pyrolysis process used. Pyrolysis oils have water contents of typically 15 to 30 wt% of the oil mass, which cannot be removed by conventional methods like distillation. Phase separation may partially occur above certain water content levels. The water content of pyrolysis oils contributes to their low energy density, lowers the flame temperature of the oils, leads to ignition difficulties, and, when preheating the oil, can lead to premature evaporation of the oil and resultant injection difficulties. The higher heating value (HHV) of pyrolysis oils is below 26 MJ/kg (compared to 42 to 45 MJ/kg for conventional petroleum fuel oils). In contrast to petroleum oils, which are nonpolar and in which water is insoluble, biomass oils are highly polar and can readily absorb over 35% water (Demirbas 2007). Table 3.5 shows the fuel properties of diesel, biodiesel, and biomass pyrolysis oil.

The bio-oil from wood is typically a liquid, almost black through dark red brown. The density of the liquid is about 1,200 kg/m^3, which is higher than that of fuel oil

Table 3.5 Fuel properties of diesel, biodiesel, and biomass pyrolysis oil

Property	Test method	ASTM D975 (Diesel)	ASTM D6751 (biodiesel, B100)	Pyrolysis oil (bio-oil)
Flash point	D 93	325 K min	403 K	–
Water and sediment	D 2709	0.05 max %vol	0.05 max %vol	0.01–0.04
Kinematic viscosity (at 313 K)	D 445	1.3–4.1 mm^2/s	1.9–6.0 mm^2/s	25–1,000
Sulfated ash	D 874	–	0.02 max %wt	–
Ash	D 482	0.01 max %wt	–	0.05–0.01 %wt
Sulfur	D 5453	0.05 max %wt	–	–
Sulfur	D 2622/129	–	0.05 max %wt	0.001–0.02 %wt
Copper strip corrosion	D 130	No 3 max	No 3 max	–
Cetane number	D 613	40 min	47 min	–
Aromaticity	D 1319	–	35 max %vol	–
Carbon residue	D 4530	–	0.05 max %mass	0.001–0.02 %wt
Carbon residue	D 524	0.35 max %mass	–	–
Distillation temp (90% volume recycle)	D 1160	555 K min –611 K max	–	–

3.6 Biogas 61

and significantly higher than that of the original biomass. Bio-oils typically have water contents of 14 to 33%wt, which cannot be removed by conventional methods like distillation. Phase separation may occur above certain water content levels. The higher heating value (HHV) is below 27 MJ/kg (compared to 43 to 46 MJ/kg for conventional fuel oils).

The bio-oil formed at 725 K contains high concentrations of compounds such as acetic acid, 1-hydroxy-2-butanone, 1-hydroxy-2-propanone, methanol, 2,6-dimethoxyphenol, 4-methyl-2,6-dimetoxyphenol, 2-cyclopenten-1-one, etc. A significant characteristic of bio-oils is the high percentage of alkylated compounds, especially methyl derivatives. As the temperature increases, some of these compounds are transformed via hydrolysis. The formation of unsaturated compounds from biomass materials generally involves a variety of reaction pathways such as dehydration, cyclization, Diels–Alder cycloaddition reactions, and ring rearrangement. For example, 2,5-hexanedione can undergo cyclization under hydrothermal conditions to produce 3-methyl-2-cyclopenten-1-one with very high selectivity of up to 81% (An *et al.* 1997).

3.6 Biogas

Biogas can be obtained from several sources. There are a number of processes for converting biomass into gaseous fuels such as methane or hydrogen. One uses plant and animal wastes in a fermentation process leading to biogas from which the desired fuels can be isolated. This technology is established and in widespread use for waste treatment. Anaerobic digestion of biowastes occurs in the absence of air, the resulting gas, called biogas, is a mixture consisting mainly of methane and carbon dioxide. Biogas is a valuable fuel that is produced in digesters filled with feedstock like dung or sewage. The digestion is allowed to continue for a period of 10 d to a few weeks. A second process uses algae and bacteria that have been genetically modified to produce hydrogen directly instead of the conventional biological energy carriers. Finally, high-temperature gasification supplies a crude gas, which may be transformed into hydrogen by a second reaction step. This process may offer the highest overall efficiency.

Anaerobic digestion (AD) is a bacterial fermentation process that is sometimes employed in wastewater treatment for sludge degradation and stabilization. This is also the principal process occurring in the decomposition of food wastes and other biomass in landfills. The AD operates without free oxygen and results in biogas containing mostly CH_4 and CO_2 but frequently carrying other substances such as moisture, hydrogen sulfide (H_2S), and particulate matter that are generally removed prior to use of the biogas. The AD is a biochemical process for converting biogenic solid waste into a stable, humuslike product. Aerobic conversion uses air or oxygen to support the metabolism of the aerobic microorganisms degrading the substrate. Aerobic conversion includes composting and activated sludge wastewater treatment processes. Composting produces useful materials, such as mulch, soil additives and amendments, and fertilizers.

AD is known to occur over a wide temperature range from 283 to 344 K. It requires attention to the nutritional needs of the facultative and methanogenic bacteria degrading the waste substrates as well as maintenance of reasonable temperatures for those bacteria. The carbon/nitrogen (C/N) ratio of the feedstock is especially important. Biogas can be used after appropriate gas cleanup as a fuel for engines, gas turbines, fuel cells, boilers, industrial heaters, other processes, and the manufacture of chemicals. AD is also being explored as a route for direct conversion into hydrogen.

Cellulose and hemicelluloses can be hydrolyzed into simple sugars and amino acids that are consumed and transformed by fermentive bacteria. Lignin is refractory to hydrolysis and generally exits the process undigested. In fact, lignin may be the most recalcitrant naturally produced organic chemical. Lignin polymers are cross-linked carbohydrate structures with molecular weights on the order of 10,000 atomic mass units. As such, lignin can bind with or encapsulate cellulose, making that cellulose unavailable to hydrolysis and digestion. Lignin degradation (or delignification of lignocellulosics) in nature is due principally to aerobic filamentous fungi that decompose the lignin in order to gain access to the cellulose and hemicelluloses.

For anaerobic systems, methane gas is an important product. Depending on the type and nature of the biological components, different yields can be obtained for different biodegradable wastes. For pure cellulose, for example, the biogas product is 50% methane and 50% carbon dioxide. Mixed waste feedstocks yield biogas with methane concentrations of 40 to 60% (by volume). Fats and oils can yield biogas with 70% methane content.

Anaerobic digestion functions over a wide temperature range from the so-called psychrophilic temperature near 283 K to extreme thermophilic temperatures above 344 K. The temperature of the reaction has a very strong influence on the anaerobic activity, but there are two optimal temperature ranges in which microbial activity and biogas production rate are highest, the so-called mesophilic and thermophilic ranges. The mesophilic regime is associated with temperatures of about 308 K, a thermophilic regime of about 328 K. Operation at thermophilic temperature allows for shorter retention time and a higher biogas production rate; however, maintaining the high temperature generally requires an outside heat source because anaerobic bacteria do not generate sufficient heat. Aerobic composting can achieve relatively high temperatures (up to 344 K) without heat addition because reaction rates for aerobic systems are much higher than those for anaerobic systems. If heat is not conducted away from the hot center of a compost pile, then this could cause thermochemical reactions that might lead to spontaneous combustion if sufficient oxygen reaches the hot areas. Managed compost operations use aeration to provide oxygen to the bacteria but also to transport heat out of the pile. The molecular structure of the biodegradable portion of the waste that contains proteins and carbohydrates is first broken down through hydrolysis. The lipids are converted to volatile fatty acids and amino acids. Carbohydrates and proteins are hydrolyzed to sugars and amino acids. In acetogenesis, acid forming bacteria use these byproducts to generate intermediary products such as propionate and butyrate. Further micro-

3.6 Biogas

bial action results in the degradation of these intermediary products into hydrogen and acetate. Methanogenic bacteria consume the hydrogen and acetate to produce methane and carbon dioxide.

After the first 6 d of digestion, methane production from manure increases exponentially, after 16 d it reaches a plateau value, and at the end of day 20, the digestion reaches the stationary phase. For wheat straw and mixtures of manure and straw the rates of digestion are lower than that of manure.

The maximum daily biogas productions are between 4 and 6 d. During a 30-d digestion period, approx. 80 to 85% of the biogas is produced in the first 15 to 18 d. This implies that the digester retention time can be designed to 15 to 18 d instead of 30 d. For the first 3 d, methane yield is almost 0% and carbon dioxide generation is almost 100%. In this period, digestion occurs as fermentation to carbon dioxide. The yields of methane and carbon dioxide gases are 50/50 at day 11. At the end of day 20, digestion reaches the stationary phase. The methane content of the biogas is in the range of 73 to 79% for the runs, the remainder being principally carbon dioxide. During digestion, the volatile fatty acid concentration is lower and the pH higher. The pH of the slurry with manure increases from 6.4 initially to 6.9 to 7.0 at the maximum methane production rate. The pH of the slurry with wheat straw is around 7.0 to 7.1 at the maximum methane production rate.

The first methane digester plant was built at a leper colony in Bombay, India, in 1859 (Meynell 1976). Most of the biogas plants utilize animal dung or sewage. A schematic of biogas plant utilizing cow dung is illustrated in Figure 3.5 (Balat 2008). AD is a commercially proven technology and is widely used for treating high-moisture-content organic wastes including +80 to 90% moisture. Biogas can be used directly in spark-ignition gas engines (SIGEs) and gas turbines. Used as a fuel in SIGEs to produce electricity only, the overall conversion efficiency from biomass to electricity is about 10 to 16% (Demirbas 2006).

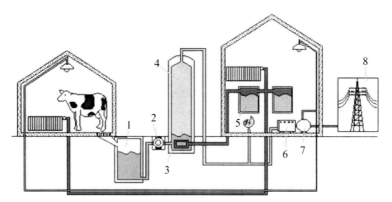

Figure 3.5 Biogas plant utilizing cow dung: 1. compost storage, 2. pump, 3. internal heater, 4. digester, 5. combustor, 6–8. power generator

3.7 Fischer–Tropsh Liquids from Biorenewable Feedstocks

Gasification followed by FTS is currently the most promising method for upgrading low-value coal and biomass to high-value liquid fuels and chemicals. The total biomass produced each year as waste material from agriculture and forest operations could be converted into roughly 40 billion gal/year of liquid fuels (roughly 25% of the current US gasoline usage).

Tijmensen *et al.* (2002) review the technical feasibility and economics of BIG-FT process and also point out the key R&D issues involved in the commercialization of this process. Boerrigter and den Uil (2002) give a similar review identifying a potential BIG-FT process configuration. The FTS for the production of liquid hydrocarbons from coal-based synthesis gas has been the subject of renewed interest for conversion of coal and natural gas to liquid fuels (Jin and Datye 2000).

Gasification is a complex thermochemical process that consists of a number of elementary chemical reactions, beginning with the partial oxidation of a biomass fuel with a gasifying agent, usually air, oxygen, or steam. The chemical reactions involved in gasification include many reactants and many possible reaction paths. The yield from the process is a product gas from thermal decomposition composed of CO, CO_2, H_2O, H_2, CH_4, other gaseous hydrocarbons, tars, char, inorganic constituents, and ash. The gas composition of product from biomass gasification depends heavily on the gasification process, the gasifying agent, and the feedstock composition. A generalized reaction describing biomass gasification is as follows:

$$\text{Biomass} + O_2 \rightarrow CO, CO_2, H_2O, H_2, CH_4 + \text{other (CHs)} + \text{tar} + \text{char} + \text{ash} \quad (3.1)$$

The relative amount of CO, CO_2, H_2O, H_2, and (CHs) depends on the stoichiometry of the gasification process. If air is used as the gasifying agent, then roughly half of the product gas is N_2.

Most biomass gasification systems utilize air or oxygen in partial oxidation or combustion processes. These processes suffer from low thermal efficiencies and low Btu gas because of the energy required to evaporate the moisture typically inherent in the biomass and the oxidation of a portion of the feedstock to produce this energy.

Syngas (a mixture of carbon monoxide and hydrogen) produced by gasification of fossil fuels or biomass can be converted into a large number of organic compounds that are useful as chemical feedstocks, fuels, and solvents. Many conversion technologies were developed for coal gasification but process economics have resulted in a shift to natural-gas-derived syngas. These conversion technologies successively apply similarly to biomass-derived biosyngas. Franz Fischer and Hans Tropsch first studied the conversion of syngas into larger, useful organic compounds in 1923 (Spath and Mann 2000).

The reasons for using biofuels are manifold and include energy security, environmental concerns, foreign exchange savings, and socioeconomic issues related to the rural sector. Catalytic conversion will be a primary tool for industry to produce valuable fuels, chemicals, and materials from biomass platform chemicals. Catalytic

3.7 Fischer–Tropsch Liquids from Biorenewable Feedstocks

conversion of biomass is best developed for synthesis gas, or syngas. Economic considerations dictate that the current production of liquid fuels from syngas translates into the use of natural gas as the hydrocarbon source. Biomass is the only renewable that can meet our demand for carbon-based liquid fuels and chemicals. Biofuels as well as green motor fuels produced from biomass by FTS are the most modern biomass-based transportation fuels. Green motor fuels are the renewable replacement for petroleum-based diesel. Biomass energy conversion facilities are important for obtaining bio-oil by pyrolysis. The main aim of FTS is the synthesis of long-chain hydrocarbons from a CO-H_2 gas mixture. The products from FTS are mainly aliphatic straight-chain hydrocarbons (C_xH_y). Besides the C_xH_y, branched hydrocarbons, unsaturated hydrocarbons, and primary alcohols are also formed in minor quantities. The FTS process is a process capable of producing liquid hydrocarbon fuels from biosyngas. The large hydrocarbons can be hydrocracked to form mainly diesel of excellent quality. The process for producing liquid fuels from biomass, which integrates biomass gasification with FTS, converts a renewable feedstock into a clean fuel.

FTS is a process for producing mainly straight-chain hydrocarbons from a syngas rich in CO and H_2. Catalysts are usually employed. Typical operating conditions for FTS are temperatures of 475 to 625 K and very high pressure depending on the desired products. The product range includes light hydrocarbons such as methane (CH_4) and ethane (C_2H_6), propane (C_3H_8), butane (C_4H_{10}), gasoline (C_5–C_{12}), diesel (C_{13}–C_{22}), and light waxes (C_{23}–C_{33}). The distribution of the products depends on the catalyst and the process conditions (temperature, pressure, and residence time). The syngas must have very low tar and particulate matter content.

The literature dealing with the actual conversion of biosyngas to fuels using FTS is smaller. Jun et al. (2004) report experimental results of FTS carried out using a model biosyngas. In his review on biofuels, Demirbas (2007) considers FTS using biosyngas as an emerging alternative.

FTS was established in 1923 by German scientists Franz Fischer and Hans Tropsch. It is described by the following set of equations (Schulz 1999):

$$nCO + (n + m/2)H_2 \rightarrow C_nH_m + nH_2O \tag{3.2}$$

where n is the average length of the hydrocarbon chain and m is the number of hydrogen atoms per carbon. All reactions are exothermic and the product is a mixture of different hydrocarbons where paraffin and olefins are the main constituents.

In FTS one mole of CO reacts with two moles of H_2 in the presence of a cobalt (Co)-based catalyst to yield a hydrocarbon chain extension ($-CH_2-$). The reaction of synthesis is exothermic ($\Delta H = -165$ kJ/mol):

$$CO + 2H_2 \rightarrow -CH_2- + H_2O \qquad \Delta H = -165 \, \text{kJ/mol} \tag{3.3}$$

The $-CH_2-$ is a building block for longer hydrocarbons. A main characteristic regarding the performance of FTS is the liquid selectivity of the process (Stelmachowski and Nowicki 2003). When iron (Fe)-based catalysts are used in WGS reaction activity, the water produced in Reaction 3.2 can react with CO to form additional H_2. The reaction of synthesis is exothermic ($\Delta H = -204$ kJ/mol). In this case a minimal H_2/CO ratio of 0.7 is required:

$$2CO + H_2 \rightarrow -CH_2- + CO_2 \qquad \Delta H = -204 \text{ kJ/mol} \qquad (3.4)$$

The kind and quantity of liquid product obtained in FTS is determined by the reaction temperature, pressure and residence time, type of reactor, and catalyst used. Fe catalysts have a higher tolerance for sulfur, are cheaper, and produce more olefin products and alcohols. However, the lifetime of Fe catalysts is short and in commercial installations generally limited to 8 weeks (Davis 2002). Bulk Fe catalysts are the catalysts of choice for converting low H_2/CO ratio syngas produced by gasification of biomass or coal to fuels via FTS. These relatively low-cost catalysts have low methane selectivity and high WGS activity. However, development of a bulk Fe FTS catalyst that combines high FT activity, low methane selectivity, high attrition resistance, and long-term stability is still elusive and presents a widely recognized barrier to the commercial deployment of FTS for biomass conversion. The critical property determining the activity and deactivation of Fe catalysts for FTS appears not to be Fe in the metallic state but the carburized Fe surface.

The design of a biomass gasifier integrated with an FTS reactor must be aimed at achieving a high yield of liquid hydrocarbons. For the gasifier, it is important to avoid methane formation as much as possible and convert all carbon in the biomass to mainly carbon monoxide and carbon dioxide (Prins *et al.* 2004). Gas cleaning is an important process before FTS and is even more important for the integration of a biomass gasifier and a catalytic reactor. To avoid the poisoning of the FTS catalyst, tar, hydrogen sulfide, carbonyl sulfide, ammonia, hydrogen cyanide, alkali, and dust particles must be removed thoroughly (Stelmachowski and Nowicki 2003).

FTS has been widely investigated for more than 70 years, and Fe and Co are typical catalysts. Co-based catalysts are preferred because their productivity is better than that of Fe catalysts thanks to their high activity, selectivity for linear hydrocarbons, and low activity for the competing WGS reaction.

There has been increasing interest in the effect of water on Co FTS catalysts in recent years. Water is produced in large amounts with Co catalysts since one water molecule is produced for each C atom added to a growing hydrocarbon chain and due to the low WGS activity of Co. The presence of water during FTS may affect the synthesis rate reversibly as reported for titania-supported catalysts and the deactivation rate as reported for alumina-supported catalysts; water also has a significant effect on the selectivity for Co catalysts on different supports. The effect on the rate and on deactivation appears to depend on the catalyst system studied, while the main trends in the effect on selectivity appear to be more consistent for different supported Co systems. There are, however, also some differences in the selectivity effects observed. The present study deals mainly with the effect of water on the selectivity of alumina-supported Co catalysts, but some data on the activity change will also be reported. The results will be compared with those for other supported Co systems reported in the literature.

3.7 Fischer–Tropsch Liquids from Biorenewable Feedstocks

The activity and selectivity of supported Co FTS catalysts depends on both the number of Co surface atoms and on their density within support particles, as well as on transport limitations that restrict access to these sites. Catalyst preparation variables available to modify these properties include Co precursor type and loading level, support composition and structure, pretreatment procedures, and the presence of promoters or additives. Secondary reactions can strongly influence product selectivity. For example, the presence of acid sites can lead to the useful formation of branched paraffins directly during the FTS step. However, product water not only oxidizes Co sites, making them inactive for additional turnovers, but it can inhibit secondary isomerization reactions on any acid sites intentionally placed in FTS reactors.

Fe catalysts used commercially in FTS for the past five decades (Dry 2004) have several advantages: (1) lower cost relative to Co and ruthenium catalysts, (2) high WGS activity allowing utilization of syngas feeds of relatively low hydrogen content such as those produced by gasification of coal and biomass, (3) relatively high activity for the production of liquid and waxy hydrocarbons readily refined to gasoline and diesel fuels, and (4) high selectivity for olefinic C_2–C_6 hydrocarbons used as chemical feedstocks. The typical catalyst used in fixed-bed reactors is an unsupported Fe/Cu/K catalyst prepared by precipitation. While having the aforementioned advantages, this catalyst (1) deactivates irreversibly over a period of months to a few years by sintering, oxidation, formation of inactive surface carbons, and transformation of active carbide phases to inactive carbide phases and (2) undergoes attrition at unacceptably high rates in the otherwise highly efficient, economical slurry bubble-column reactor.

It is well known that the addition of alkali to iron causes an increase in both the 1-alkene selectivity and the average carbon number of produced hydrocarbons. While the promoter effects on iron have been thoroughly studied, few and, at first glance, contradictory results are available for Co catalysts. In order to complete the experimental data, the carbon number distributions are analyzed for products obtained in a fixed-bed reactor under steady-state conditions. Precipitated Fe and Co catalysts with and without K_2CO_3 were used.

Activated carbon (AC) is a high-surface-area support with the very unique property that its textural and surface chemical properties can be changed by an easy treatment like oxidation, and these changes affect the properties of the resultant catalysts prepared with AC.

Fe catalysts have a higher tolerance for sulfur, are cheaper, and produce more olefin products and alcohols. However, the lifetime of Fe catalysts is short and in commercial installations generally limited to 8 weeks. Co catalysts have the advantage of a higher conversion rate and a longer life (over 5 years). Co catalysts are in general more reactive for hydrogenation and therefore produce fewer unsaturated hydrocarbons and alcohols compared to Fe catalysts.

Low-temperature Fischer–Tropsch (LTFT) reactors, either multitubular fixed-bed (MTFBR) or slurry reactor (SR) operating at approx. 25 bar and 495 to 525 K, use a precipitated FE catalyst. High-temperature Fischer–Tropsch (HTFT) fluidized bed reactors, either fixed (SAS) or circulating (Synthol) operating at approx. 25 bar and

575 to 595 K, use a fused Fe catalyst. In their experiments with Fe for CO_2 hydrogenation, Riedel *et al.* (1999) found that alumina was the best support and potassium acted as a powerful promoter. Copper was added to the catalyst to enable its easy reduction. They report that the hydrocarbon distribution from the H_2/CO_2 and H_2/CO syngas is the same, but the reaction rate for CO_2 syngas was about 43% lower than that of the CO-rich syngas.

The Al_2O_3/SiO_2 ratio has a significant influence on Fe-based catalyst activity and selectivity in the process of FTS. Product selectivities also change significantly with different Al_2O_3/SiO_2 ratios. The selectivity of low-molecular-weight hydrocarbons increases and the olefin-to-paraffin ratio in the products shows a monotonic decrease with an increasing Al_2O_3/SiO_2 ratio. Table 3.6 shows the effects of the Al_2O_3/SiO_2 ratio on hydrocarbon selectivity (Jothimurugesan *et al.* 2000). Recently, Jun *et al.* (2004) studied FTS over Al_2O_3- and SiO_2-supported Fe-based catalysts from biomass-derived syngas. They found that Al_2O_3 as a structural promoter facilitated the better dispersion of copper and potassium and gave much higher FTS activity. Table 3.7 shows properties of FT diesel and No. 2 diesel fuels.

Biosyngas consists mainly of H_2, CO, CO_2, and CH_4. FTS has been carried out using a $CO/CO_2/H_2/Ar$ (11/32/52/5 vol.%) mixture as a model for biosyngas on co-precipitated Fe/Cu/K, Fe/Cu/Si/K, and Fe/Cu/Al/K catalysts in a fixed-bed reactor. Some performances of catalysts that depended on the syngas composition have also been presented (Jun *et al.* 2004). The kinetic model predicting product distribution is taken from Wang *et al.* (2003) for an industrial Fe-Cu-K catalyst.

Table 3.6 Effects of Al_2O_3/SiO_2 ratio on hydrocarbon selectivity

Hydrocarbon selectivities (wt%)	100Fe/6Cu/ 5K/25SiO$_2$ 22SiO$_2$	100Fe/6Cu/ 5K/3Al$_2$O$_3$/ 20SiO$_2$	100Fe/6Cu/ 5K/5Al$_2$O$_3$/ 18SiO$_2$	100Fe/6Cu/ 5K/7Al$_2$O$_3$/ 15SiO$_2$	100Fe/6Cu/ 5K/10Al$_2$O$_3$/	100Fe/6Cu/ 5K/25Al$_2$O$_3$
CH_4	6.3	8.7	10.4	10.7	14.3	17.3
C_{2-4}	24.5	27.8	30.8	29.9	33.4	46.5
C_{5-11}	26.8	27.6	32.2	33.9	40.0	31.0
C_{12-18}	21.9	21.2	15.8	15.0	6.0	4.9
C_{19+}	20.5	14.4	11.0	10.6	6.1	0.4

Reaction condition: 523 K, 2.0 MPa, H_2/CO = 2.0, and Gas stream velocity: 2,000/h.

Table 3.7 Properties of Fischer–Tropsch (FT) diesel and No. 2 diesel fuels

Property	FT diesel	No. 2 petroleum diesel
Density, g/cm^3	0.7836	0.8320
Higher heating value, MJ/kg	47.1	46.2
Aromatics, %	0–0.1	8–16
Cetane number	76–80	40–44
Sulfur content, ppm	0–0.1	25–125

FTS for the production of transportation fuels and other chemicals from synthesis has attracted much attention due to the pressure from the oil supply. Interest in the use of Fe-based catalysts stems from its relatively low cost and excellent WGS reaction activity, which helps to make up the deficit of H_2 in the syngas from coal gasification (Wu *et al.* 2004; Jothimurugesan *et al.* 2000; Jun *et al.* 2004). Riedel *et al.* (1999) have studied the hydrogenation of CO_2 over both these catalysts. In the absence of any WGS reaction promoter like Mn, CO_2 merely behaves as a diluting gas as it is neither strongly adsorbed nor hydrogenated on Co catalysts. When Mn is added to Co catalysts, reverse WGS is possible. The FT chain growth on Co occurs due to strongly adsorbed CO on the surface. With a low partial pressure of CO, these inhibitions are removed and the regime moves from an FT to a methanation regime, yielding more CH_4. It was observed that even when the r-WGS reaction was fast, the attainable CO concentration was not sufficient to attain an FT regime. It was hence concluded that CO_2 hydrogenation is not possible even with a hybrid Co catalyst containing a shift catalyst like Mn.

3.8 Biohydrogen

The combustion products of hydrogen when it is burned completely with air consist of water, oxygen, and nitrogen. However, it has been suggested that hydrogen is too valuable to burn. Laboratory tests conducted on internal combustion engines burning hydrogen demonstrate good performance (Berry *et al.* 1996). In comparison with an engine burning gasoline, the emission of nitrogen oxides is far less for the engine-fueled hydrogen. The product of hydrogen combustion with air is water vapor and negligible pollution when the peak temperature is limited. Some oxides of nitrogen (NO_x) are formed at very high combustion temperatures ($<2,300\,K$); fortunately, the autoignition temperature of hydrogen is only $858\,K$.

Hydrogen has good properties as a fuel for internal combustion engines in automobiles. Some of the characteristic properties of a hydrogen–air mixture that can definitely influence engine design and performance are low ignition energy, low density, wide range of ignition limits, high diffusion speed, and high flame speed (Plass Jr. *et al.* 1990).

The main disadvantage of using hydrogen as a fuel for automobiles is the huge on-board storage tanks that are required because of hydrogen's extremely low density. Hydrogen may be stored on board a vehicle as compressed gas in ultra-high-pressure vessels, as a liquid in cryogenic containers, or as a gas bound with certain metals in metal hydrides.

Hydrogen is one of the most promising alternative energy technologies. Hydrogen can be generated in a number of ways, such as electrochemical processes, thermochemical processes, photochemical processes, photocatalytic processes, or photoelectrochemical processes (Momirlan and Veziroglu 1999, 2002). Biohydrogen production by anaerobic fermentation of renewable organic waste sources has been found to be a promising method for the recovery of bioenergy (Han and Shin 2004).

In this method, anaerobic bacteria use organic substances as the sole source of electrons and energy, converting them into hydrogen.

The use of hydrogen as a fuel for transportation and stationary applications is receiving much favorable attention as a technical and policy issue (Cherry 2004). Hydrogen gas is being explored for use in combustion engines and fuel-cell electric vehicles. It is a gas at normal temperatures and pressures, which presents greater transportation and storage hurdles than exist for liquid fuels. Several hydrogen technologies are under development; the most promising of these is the fuel cell. Fuel cells use hydrogen, oxygen, a catalyst, and an electrolytic solution to produce energy in the form of heat and electricity.

3.9 Other Liquid Biofuels

3.9.1 Glycerol-based Fuel Oxygenates for Biodiesel and Diesel Fuel Blends

Glycerol (1,2,3-propanetriol or glycerine) is a trihydric alcohol. It is a colorless, odorless, sweet-tasting, syrupy liquid. It melts at 291 K, boils with decomposition at 563 K, and is miscible with water and ethanol (Perry and Green 1997). The chemical formula for glycerol is $OH–CH_2–CH(OH)–CH_2–OH$. Glycerol is present in the form of its esters (triglycerides) in vegetable oils and animal fats.

Glycerol is a byproduct obtained during the production of biodiesel. As biodiesel production is increasing exponentially, the crude glycerol generated from the transesterification of vegetables oils has also been generated in large quantities (Pachauri and He 2006). With the increasing production of biodiesel a glut of glycerol has been created, causing market prices to plummet. This situation warrants finding alternative uses for glycerol. Glycerol is directly produced with high purity levels (at least 98%) by biodiesel plants (Ma and Hanna 1999; Bournay et al. 2005). Research efforts to find new applications of glycerol as a low-cost feedstock for functional derivatives have led to the introduction of a number of selective processes for converting glycerol into commercially valued products (Pagliaro et al. 2007). The principal byproduct of biodiesel production is crude glycerol, which is about 10 wt% of vegetable oil. For every 9 kg of biodiesel produced, about 1 kg of a crude glycerol byproduct is formed (Dasari et al. 2005).

Oxygenated compounds such as methyl tertiary butyl ether (MTBE) are used as valuable additives as a result of their antidetonant and octane-improving properties. In this respect, glycerol tertiary butyl ether is an excellent additive with a large potential for diesel and biodiesel reformulation.

Glycerol can be converted into higher-value products. The products are 1,3-propanediol, 1,2-propanediol, dihydroxyacetones, hydrogen, polyglycerols, succinic acid, and polyesters. The main glycerol-based oxygenates are 1,3-propanediol, 1,2-propanediol, propanol, glycerol tert-butyl ethers, ethylene glycol, and propylene glycol.

3.9.2 P-series Fuels

A P-series fuel is a unique blend of liquefied petroleum gas liquids, ethanol, hydrocarbons, and methyltetrahydrofuran (MeTHF). P-series fuels are blends of ethanol, MeTHF, and pentanes, with butane added for blends that would be used in severe cold-weather conditions to meet cold start requirements. P-series fuels are made primarily from biorenewable resources and provide significant emissions benefits over reformulated gasoline. A P-series fuel can be mixed with gasoline in any proportion and used in FFVs. P-series fuels are clear, colorless, 89- to 93-octane liquid blends that are formulated for use in FFVs. Like gasoline, low-vapor-pressure formulations are produced to prevent excessive evaporation during summer and high-vapor-pressure formulations are used for easy starting in the winter. P-series fuels are at least 60% nonpetroleum. They also have many environmental benefits. Because a majority of the components that make up P-series fuels come from domestically produced renewable resources, this alternative fuel promotes both energy security and environmental quality. P-series fuels could be 96% derived from domestic resources and could reduce fossil energy use by 49 to 57% and petroleum use by 80% relative to gasoline. Greenhouse gas emissions from the production and use of P-series fuels are substantially better than those from gasoline. Each unit of P-series fuel emits approx. 50% less carbon dioxide, 35% less hydrocarbons, and 15% less carbon monoxide than gasoline. It also has 40% less ozone-forming potential.

3.9.3 Dimethyl Ether

Dimethyl ether (DME or CH_3-O-CH_3), is a new fuel that has attracted much attention recently. Today DME is made from natural gas, but DME can also be produced by gasifying biomass. DME can be stored in liquid form at 5 to 10 bars pressure at normal temperature. A major advantage of DME is its naturally high cetane number, which means that self-ignition is easier. The high cetane rating makes DME most suitable for use in diesel engines, which implies that the high level of efficiency of the diesel engine is retained when using DME. The energy content of DME is lower than in diesel.

DME can be produced effectively from biosyngas in a single-stage, liquid-phase (LPDME) process. The origin of syngas includes a wide spectrum of feedstocks such as coal, natural gas, biomass, and others. Nontoxic, high-density, liquid DME fuel can be easily stored at modest pressures. The production of DME is very similar to that of methanol. DME conversion to hydrocarbons, lower olefins in particular, has been studied using ZSM-5 catalysts with varying SiO_2/Al_2O_3 ratios, whereas the DME carbonization reaction to produce methyl acetate has been studied over a variety of group VIII metal-substituted heteropolyacid catalysts.

3.9.4 Other Bio-oxygenated Liquid Fuels

Methanol and ethanol are not the only transportation fuels that might be made from wood. A number of possibilities exist for producing alternatives. The most promising bio-oxygenated fuels, and closest to being competitive in current markets without subsidy, are ethanol, methanol, ethyl-tert-butyl ether, and anti-methyl-tert-butyl ether. Other candidates include isopropyl alcohol, sec-butyl alcohol, tert-butyl alcohol, mixed alcohols, and tert-amylmethyl ether.

Another possibility for bio-oxygenated fuels is methanol. Methanol could conceivably be made from grain, but its most common source is natural gas. The use of natural gas is better for reducing carbon dioxide production in comparison to other fossil fuels, but the use of renewable fuels instead of natural gas would be better still. It can be made from coal or wood with more difficulty and lower efficiency than from natural gas. The cost of making methanol from natural gas is around US$ 0.40 per gallon. It could probably be sold as a motor fuel for about US$ 0.60 to $ 0.70 per gallon. This would be equivalent to gasoline selling at about US$ 0.92 to $ 1.03 per gallon. Methanol was once produced from wood as a byproduct of charcoal manufacture, but overall yields were low. To produce methanol from wood with a significantly higher yield would require production of synthesis gas in a process similar to that used for production of methanol from coal. Such processes for gasifying wood are less fully developed than the two-stage hydrolysis process for the production of ethanol.

A high octane rating is characteristic of all oxygenated fuels, including ethanol, methanol, ethyl-tert-butyl ether (ETBE), and MTBE. MTBE is made by reacting isobutylene with methanol. ETBE is made by using ethanol instead of methanol. Thus either ethanol or methanol from either grain or wood could be a factor in making tert-butyl ether octane enhancers. The characteristics of ethers are generally closer to those of gasolines than those of alcohols. Ethers are benign in their effect on fuel system materials and are miscible in gasoline; therefore, they are not subject to phase separation in the presence of water, as are methanol and ethanol.

References

An, J., Bagnell, L., Cablewski, T., Strauss, C. R., Trainor, R. W. 1997. Applications of high-temperature aqueous media for synthetic organic reactions. J Org Chem 62:2505–2511.

Balat, M. 2008. Progress in biogas production processes. Energy Edu Sci Technol 22:15–35.

Balat, M. 2009. New biofuel production technologies. Energy Educ Sci Technol Part A 22:147–161.

Bender, M. 1999. Economic feasibility review for community-scale farmer cooperatives for biodiesel. Bioresour Technol 70:81–87.

Berry, G. D., Pasternak, A. D., Rambach, G. D., Smith, J. R., Schock, R. N. 1996. Hydrogen as a future transportation fuel. Energy 21:289–303.

Boerrigter, H., den Uil, H. 2002. Green diesel from biomass via FTS: new insights in gas cleaning and process design. Pyrolysis and Gasification of Biomass and Waste, Expert Meeting, Strasbourg, France (30 Sep.–1 Oct.).

References

Bournay, L., Casanave, D., Delfort, B., Hillion, G., Chodorge, J. A. 2005. New heterogeneous process for biodiesel production: a way to improve the quality and the value of the crude glycerin produced by biodiesel plants. Catal Today 106:190–192.

Cherry, R. S. 2004. A hydrogen utopia? Int J Hydrogen Energy 29:125–129.

Dasari, M. A., Kiatsimkul, P. P., Sutterlin, W. R., Suppes, G. J. 2005. Low-pressure hydrogenolysis of glycerol to propylene glycol. Appl Catal A 281:225–231.

Davis, B. H. 2002. Overview of reactors for liquid phase Fischer–Tropsch synthesis. Catal Today 71:249–300.

Demirbas, A. 2006. Biogas potential of manure and straw mixtures. Energy Sour A 28:71–78.

Demirbas, A. 2007. Progress and recent trends in biofuels. Prog Energy Combus Sci 33:1–18.

Demirbas, A. 2008. Recent progress in biorenewable feedstocks. Energy Educ Sci Technol 22:69–95.

Dincer, K. 2008. Lower emissions from biodiesel combustion. Energy Sour A 30:963–968.

Dry, M. E. 2002. The Fischer–Tropsch process: 1950–2000. Catal Today 71:227–241.

Dry, M. E. 2004. Present and future applications of the Fischer–Tropsch process. Appl Catal A 276:1–3.

EBB. 2009. EU: Biodiesel industry expanding use of oilseeds. European Biodiesel Board, Brussels.

Goldemberg, J. 2000. World Energy Assessment, United Nations Development Programme, New York.

Haas, M. J., McAloon, A. J., Yee, W. J., Foglia, T. A. 2006. A process model to estimate biodiesel production costs. Bioresour Technol 97:671–678.

Han, S.-K., Shin, H.-S. 2004. Biohydrogen production by anaerobic fermentation of food waste. Int J Hydrogen Energy 29:569–577.

IEA. 2007. Key world energy statistics. International Energy Agency, Paris. http://www.iea.org/Textbase/nppdf/free/2007/key_stats_2007.pdf.

Jin, Y., Datye, A. K. 2000. Phase transformations in iron Fischer–Tropsch catalysts during temperature-programmed reduction. J Catal 196:8–17.

Jothimurugesan, K., Goodwin, J. G., Santosh, S. K., Spivey, J. J. 2000. Development of Fe Fischer–Tropsch catalysts for slurry bubble column reactors. Catal Today 58:335–344.

Jun, K. W., Roh, H. S., Kim, K. S., Ryu, J. S., Lee, K. W. 2004. Catalytic investigation for Fischer–Tropsch synthesis from bio-mass derived syngas. Appl Catal A 259:221–226.

Kadiman, O. K. 2005. Crops: beyond foods. In: Proceedings of the 1st International Conference on Crop Security, Malang, Indonesia, 20–23 September 2005.

Keskin, A. 2009. Biodiesel production from free fatty acids obtained with neutralization of the crude glycerin. Energy Sour A 31:17–24.

Knothe, G., Krahl, J., Van Gerpen, J. (eds.). 2005. The Biodiesel Handbook. AOCS Press, Champaign, IL.

Knothe, G., Sharp, C. A., Ryan, T. W. 2006. Exhaust emissions of biodiesel, petrodiesel, neat methyl esters, and alkanes in a new technology engine. Energy Fuels 20:403–408.

Ma, F., Hanna, M. A. 1999. Biodiesel production: a review. Biores Technol 70:1–15.

Meynell, P.-J. 1976. Methane: Planning a Digester. Schocken, New York.

Mislavskaya, V. S., Leonow, V. E., Mislavskii, N. O., Ryzhak, I. A. 1982. Conditions of phase stability in a gasoline-methanol-cyclohexanol-water system. Soviet Chem Ind 14:270–276.

Mohan, D., Pittman Jr., C. U., Steele, P. H. 2006. Pyrolysis of wood/biomass for bio-oil: a critical review. Energy Fuels 2006:20:848–889.

Momirlan, M., Veziroglu, T. 1999. Recent directions of world hydrogen production. Renew Sust Energy Rev 3:219–231.

Momirlan, M., Veziroglu, T. 2002. Current status of hydrogen energy. Renew Sustain Energy Rev 6:141–79.

Osten, D. W., Sell, N. J. 1983. Methanol-gasoline blends: Blending agents to prevent phase separation. Fuel 62:268–270.

Pachauri, N., He, B. 2006. Value-added utilization of crude glycerol from biodiesel production: a survey of current research activities. ASABE Annual International Meeting, Portland, OR, 9–12 July 2006, pp. 1–16.

Pagliaro, M., Ciriminna, R., Kimura, H., Rossi, M., Pina, C. D. 2007. From glycerol to value-added products. Angew Chem Int Ed 46:4434–440.

Perry, R. H., Green, D. W. 1997. Perry's Chemical Engineers' Handbook. pp2–39. McGraw-Hill, New York.

Plass Jr., J. H., Barbir, F., Miller, H. P., Veziroglu, T. N. 1990. Economics of hydrogen as a fuel for surface transportation. Int J Hydrogen Energy 15:663–668.

Prins, M. J., Ptasinski, K. J, Janssen, F. J. J. G. 2004. Exergetic optimisation of a production process of Fischer–Tropsch fuels from biomass. Fuel Proc Technol 86:375–389.

Pryor, R. W., Hanna, M. A., Schinstock, J. L., Bashford, L. L. 1983. Soybean oil fuel in a small diesel engine. Trans ASAE 26:333–338.

RFA. 2009. Renewable Fuels Association (RFA). Ethanol Industry Statistics, Washington, DC, USA.

Riedel, T., Claeys, M., Schulz, H., Schaub, G., Nam, S. S., Jun, K. W., Choi, M. J., Kishan, G., Lee, K. W. 1999. Comparative study of FTS with H_2/CO and H_2/CO_2 syngas using Fe and Co catalysts. Appl Catal A 186:201–213.

Schulz, H. 1999. Short history and present trends of FT synthesis. Appl Catal A 186:1–16.

Spath, P. L., Mann, M. K. 2000. Life cycle assessment of hydrogen production via natural gas steam reforming. National Renewable Energy Laboratory, Golden, CO, TP-570–27637, November.

Stelmachowski, M., Nowicki, L. 2003. Fuel from synthesis gas–the role of process engineering. Appl Energy 74:85–93.

Tijmensen, M. J. A., Faaij, A. P. C., Hamelinck, C. N., van Hardeveld, M. R. M. 2002. Exploration of the possibilities for production of Fischer Tropsch liquids and power via biomass gasification. Biomass Bioenergy 23:129–152.

UN (United Nations). 2006. The emerging biofuels market: regulatory, trade and development implications. United Nations conference on trade and development, New York and Geneva.

Wang, Y. N., Ma, W. P., Lu, Y. J., Yang, J., Xu, Y. Y., Xiang, H. W., Li, Y. W., Zhao, Y. L., Zhang, B. J. 2003. Kinetics modeling of FT synthesis over an industrial Fe-Cu-K catalyst. Fuel 82:195–213.

Wu, B. S., Bai, L., Xiang, H. W., Li, Y. W., Zhang, Z. X., Zhong, B. 2004. An active iron catalyst containing sulfur for Fischer–Tropsch synthesis. Fuel 83:205–512.

Zhang, Y., Dub, M. A., McLean, D. D., Kates, M. 2003. Biodiesel production from waste cooking oil: 2. Economic assessment and sensitivity analysis. Biores Technol 90:229–240.

Chapter 4
Algae Technology

4.1 Introduction

Billions of years ago the Earth's atmosphere was filled with CO_2. Thus there was no life on the planet. Life on Earth started with *Cyanobacterium* and algae. These humble photosynthetic organisms sucked out the atmospheric CO_2 and started releasing oxygen. As a result, the levels of CO_2 started decreasing to such an extent that life evolved on Earth. Once again these smallest organisms are poised to save us from the threat of global warming.

Algae, like corn, soybeans, sugar cane, wood, and other plants, use photosynthesis to convert solar energy into chemical energy. They store this energy in the form of oils, carbohydrates, and proteins. The more efficient a particular plant is at converting that solar energy into chemical energy, the better it is from a biodiesel perspective, and algae are among the most photosynthetically efficient plants on earth. A 1-ha algae farm on wasteland can produce over 10 to 100 times of oil as compared to any other known source of oil crop. Algae-based technologies could provide a key tool for reducing greenhouse gas emissions from coal-fired power plants and other carbon-intensive industrial processes.

Algae range from small, single-celled organisms to multicellular organisms, some with fairly complex and differentiated forms. Algae are usually found in damp places or bodies of water and thus are common in terrestrial as well as aquatic environments. Like plants, algae require primarily three components to grow: sunlight, carbon dioxide, and water. Photosynthesis is an important biochemical process in which plants, algae, and some bacteria convert the energy of sunlight into chemical energy.

Microalgae are fast-growing beasts with a voracious appetite for carbon dioxide. They have the potential to produce more oil per acre than any other feedstock being used to make biodiesel, and they can be grown on land that's unsuitable for food crops (Demirbas 2009a).

A. Demirbas, M. Fatih Demirbas, *Algae Energy*
DOI 10.1007/978-1-84996-050-2, © Springer 2010

Algae are simple organisms that are mainly aquatic and microscopic. Microalgae are unicellular photosynthetic microorganisms living in saline or freshwater environments that convert sunlight, water, and carbon dioxide into algal biomass (Ozkurt 2009). They are categorized into four main classes: diatoms, green algae, blue-green algae, and golden algae. There are two main populations of algae: filamentous and phytoplankton. These two species, in particular phytoplankton, increase in numbers rapidly to form algae blooms (Demirbas 2009b). Like higher plants, they produce storage lipids in the form of TAGs. Many species exhibit rapid growth and high productivity, and many microalgal species can be induced to accumulate substantial quantities of lipids, often greater than 60% of their dry biomass (Sheehan *et al.* 1998). Microalgae are very efficient solar energy converters and can produce a great variety of metabolites. Humans have always tried to take advantage of these properties through algal mass culture.

To achieve environmental and economic sustainability, fuel production processes are required that are not only renewable but also capable of sequestering atmospheric CO_2. Currently, nearly all renewable energy sources target the electricity market, while fuels make up a much larger share of the global energy demand. Biofuels are therefore rapidly being developed. Second-generation microalgal systems have the advantage that they can produce a wide range of feedstocks for the production of biodiesel, bioethanol, biomethane, and biohydrogen. Biodiesel is currently produced from oil synthesized by conventional fuel crops that harvest the Sun's energy and store it as chemical energy. This presents a route for renewable and carbon-neutral fuel production. However, current supplies from oil crops and animal fats account for only approx. 0.3% of the current demand for transport fuels. Increasing biofuel production on arable land could have severe consequences for the global food supply. In contrast, producing biodiesel from algae is widely regarded as one of the most efficient ways of generating biofuels and also appears to represent the only current renewable source of oil that could meet the global demand for transport fuels (Schenk *et al.* 2008).

4.1.1 Definition of Algal Terms

Algal organisms are photosynthetic macroalgae or microalgae growing in aquatic environments. Macroalgae, or "seaweeds," are multicellular plants growing in salt or fresh water. They are often fast growing and can reach sizes of up to 60 m in length (McHugh 2003). They are classified into three broad groups based on their pigmentation: (1) brown seaweed (*Phaeophyceae*), (2) red seaweed (*Rhodophyceae*), and (3) green seaweed (*Chlorophyceae*).

Microalgae are microscopic photosynthetic organisms that are found in both marine and freshwater environments. Biologists have categorized microalgae into a variety of classes, mainly distinguished by their pigmentation, life cycle, and basic cellular structure. The three most important classes of microalgae in terms of abundance are the diatoms (*Bacillariophyceae*), the green algae (*Chlorophyceae*), and the golden algae (*Chrysophyceae*). The cyanobacteria (blue-green algae) (*Cyanophyceae*) are also referred to as microalgae. This applies, for example, to Spirulina (*Arthrospira platensis* and *A. maxima*). Diatoms are the dominant life form in phytoplankton and probably represent the largest group of biomass producers on Earth. It is estimated that more than 100,000 species exist.

Microalgae are primitive organisms with a simple cellular structure and a large surface-to-volume-body ratio, which gives them the ability to uptake large amounts of nutrients (Sheehan *et al.* 1998). The photosynthetic mechanism of microalgae is similar to land-based plants, but, due to their simple cellular structure and to the fact they are submerged in an aqueous environment where they have efficient access to water, CO_2, and other nutrients, they are generally more efficient in converting solar energy into biomass (Carlsson *et al.* 2007). The growth medium must contribute the inorganic elements that help make up the algal cell such as nitrogen, phosphorus, iron, and sometimes silicon (Grobbelaar 2004).

Microalgae can be used for bioenergy generation (biodiesel, biomethane, biohydrogen), or combined applications for biofuel production and CO_2 mitigation. Microalgae are veritable miniature biochemical factories and appear more photosynthetically efficient than terrestrial plants (Pirt 1986) and are efficient CO_2 fixers (Brown and Zeiler 1993).

The existing large-scale natural sources of algae are bogs, marshes, and swamps – salt marshes and salt lakes. Microalgae contain lipids and fatty acids as membrane components, storage products, metabolites, and sources of energy. Algae contain anywhere from 2 to 40% of lipids/oils by weight. Essential elements include nitrogen (N), phosphorus (P), iron, and, in some cases, silicon (Chisti 2007). Minimal nutritional requirements can be estimated using the approximate molecular formula of the microalgal biomass: $CO_{0.48}H_{1.83}N_{0.11}P_{0.01}$. This formula is based on data presented by Grobbelaar (2004).

The production of microalgal biodiesel requires large quantities of algal biomass. Macro- and microalgae are currently mainly used for food, in animal feed, in feed for aquaculture, and as biofertilizer. Biomass from microalgae is dried and marketed in the human health food market in the form of powders or pressed in the form of tablets. Aquatic biomass could also be used as raw material for cofiring to produce electricity, for liquid fuel (bio-oil) production via pyrolysis, or for biomethane generation through fermentation. Biomethane can be produced from marine biomass (Demirbas 2006).

4.2 Production Systems

The world production of seaweeds was some 8 million tons in 2003 (McHugh 2003). Seaweeds are used in the production of food, feed, chemicals, cosmetics, and pharmaceutical products.

Most microalgae are strictly photosynthetic, i.e., they need light and carbon dioxide as energy and carbon sources. This culture mode is usually called photoau-

78   4 Algae Technology

totrophic. Some algae species, however, are capable of growing in darkness and of using organic carbons (such as glucose or acetate) as energy and carbon sources. This culture mode is termed heterotrophic.

Microalgae cultivation using sunlight energy can be carried out in open or covered ponds or closed photobioreactors, based on tubular, flat plate, or other designs. Closed systems are much more expensive than ponds, present significant operating challenges (overheating, fouling), and, due to, among other things, gas exchange limitations, cannot be scaled up much beyond approx. $100\,m^2$ for an individual growth unit.

The concept of using microalgae as a source of fuel is older than most people realize. The idea of producing methane gas from algae was proposed in the early 1950s. Currently there are three types of industrial reactors used for algal culture: (1) photobioreactors, (2) open ponds, and (3) closed and hybrid systems.

Photobioreactors are different types of tanks or closed systems in which algae are cultivated. Open-pond systems are shallow ponds in which algae are cultivated. Nutrients can be provided through runoff water from nearby land areas or by channeling the water from sewage/water treatment plants. Technical and biological limitations of these open systems have given rise to the development of enclosed photoreactors. Microalgae cultivation using sunlight energy can be carried out in open or covered ponds or closed photobioreactors, based on tubular, flat plate, or other designs. A few open systems are presented for which particularly reliable results are available. Emphasis is then put on closed systems, which have been considered to be capital intensive and are justified only when a fine chemical is to be produced. Microalgae production in closed photobioreactors is highly expensive. Closed systems are much more expensive than ponds. However, closed systems require much less light and agricultural land to grow algae. High oil species of microalgae cultured in growth-optimized conditions of photobioreactors have the potential to yield 19,000 to 57,000 L of microalgal oil per acre per year. The yield of oil from algae is over 200 times the yield from the best-performing plant/vegetable oils (Chisti 2007).

Large-scale production of microalgal biomass generally uses continuous culture during daylight. In this method of operation, fresh culture medium is fed at a constant rate and the same quantity of microalgal broth is withdrawn continuously (Molina Grima et al. 1999). Feeding ceases during the night, but the mixing of broth must continue to prevent settling of the biomass (Molina Grima et al. 1999). As much as 25% of the biomass produced during daylight may be lost during the night due to respiration. The extent of this loss depends on the light level under which the biomass was grown, the growth temperature, and the temperature at night (Chisti 2007).

Algal cultures consist of a single or several specific strains optimized for producing the desired product. Water, necessary nutrients, and CO_2 are provided in a controlled way, while oxygen has to be removed (Carlsson et al. 2007). Algae receive sunlight either directly through the transparent container walls or via light fibers or tubes that channel it from sunlight collectors. A great amount of developmental work to optimize different photobioreactor systems for algae cultivation

4.2 Production Systems 79

has been carried out and is reviewed in Janssen *et al.* (2003), Choi *et al.* (2003), Carvalho *et al.* (2006), and Hankamer *et al.* (2007).

Bioreactors are the preferred method for scientific researchers, and recently for some newer, innovative production designs. These systems are more expensive to build and operate; however, they allow for a very controlled environment. This means that gas levels, temperature, pH, mixing, media concentration, and light can be optimized for maximum production (Chisti 2007). Unlike open ponds, bioreactors can ensure a single alga species is grown without interference or competition (Campbell 2008).

4.2.1 Harvesting Microalgae

Conventional processes used to harvest microalgae include concentration through centrifugation (Haesman *et al.* 2000), foam fractionation (Csordas and Wang 2004), flocculation (Poelman *et al.* 1997; Knuckey *et al.* 2006), membrane filtration (Rossignol *et al.* 2000), and ultrasonic separation (Bosma *et al.* 2003). Harvesting costs may contribute 20 to 30% to the total cost of algal biomass (Molina Grima *et al.* 2003). Microalgae are typically small with a diameter of 3 to 30 µm, and the culture broths may be quite dilute at less than 0.5 g/L. Thus, large volumes must be handled. The harvesting method depends on the species, on the cell density, and often also on the culture conditions (Carlsson *et al.* 2007).

Algae pressing is very similar to the techniques used to press flowers and is used widely by scientists as a means of preserving algal specimens and observing their features.

Algae can be harvested by centrifugation, flocculation, or froth flotation. Alum and ferric chloride are chemical flocculants used to harvest algae. Water that is brackish or salty requires additional chemical flocculants to induce flocculation. Harvesting by chemical is a method that is often too expensive for large operations. However, interrupting the carbon dioxide supply to an algal system can cause algae to flocculate on its own, which is called "autoflocculation." In froth flotation, the water and algae are aerated into froth and algae and then removed from the water.

The typical cell density achieved in the industrial application is between 0.3 and 0.5 g dry cell/L or 5 g dry cell/L at best, which makes harvest difficult and expensive (Wang *et al.* 2008). Two processes are involved in harvesting, bulk harvesting and thickening. Bulk harvesting is a large-scale operation separating biomass from bulk culture. It has a concentration factor of 100 to 800 times, depending on the culture and harvesting method. Bulk harvesting can be categorized into flocculation and floatation. Flocculation reduces/neutralizes the negative surface charge of microalgal cells, allowing them to aggregate into larger lumps with an efficiency of >80. The thickening process consists of either centrifugation or filtration. Centrifugation, a semicontinuous or continuous process, utilizes centrifugal force generated by the spinning of a suspension to separate and harvest algal cells.

80   4 Algae Technology

For mass cultivation of algae, optimization of algae harvesting and processing is needed. Algae properties such as algae size, cell wall sensitivity to shear force, ease of flocculation, and oil content need to be taken into consideration in the design process.

4.2.2 Photobioreactors

Photobioreactors have the ability to produce algae while performing beneficial tasks, such as scrubbing power plant flue gases or removing nutrients from wastewater (Carlsson *et al.* 2007). Photobioreactors are different types of tanks or closed systems in which algae are cultivated (Richmond 2004).

Algae biomass can play an important role in solving the problem between the production of food and that of biofuels in the near future. Microalgae appear to be the only source of renewable biodiesel that is capable of meeting the global demand for transport fuels. Microalgae are photosynthetic microorganisms that convert sunlight, water, and carbon dioxide into algal biomass (Chisti 2007).

Most algal species are obligate phototrophs and thus require light for their growth. Phototropic microalgae are most commonly grown in open ponds and photobioreactors (Patil *et al.* 2005). Open-pond cultures are economically more favorable but raise the issues of land use cost, water availability, and appropriate climatic conditions. Photobioreactors offer a closed culture environment, which is protected from direct fallout and so is relatively safe from invading microorganisms. This technology is relatively expensive compared to the open ponds because of the infrastructure costs. An ideal biomass production system should use the freely available sunlight.

Tredici (1999) has reviewed mass production in photobioreactors. Many different designs of photobioreactor have been developed, but a tubular photobioreactor seems to be most satisfactory for producing algal biomass on the scale needed for biofuel production. Closed, controlled, indoor algal photobioreactors driven by artificial light are already economical for special high-value products such as pharmaceuticals, which can be combined with the production of biodiesel to reduce the cost (Patil *et al.* 2008).

Photobioreactors have higher efficiency and biomass concentration (2 to 5 g/L), shorter harvest time (2 to 4 weeks), and higher surface-to-volume ratio (25 to 125/m) than open ponds (Lee 2001; Wang *et al.* 2008). Closed systems consist of numerous designs: tubular, flat-plated, rectangular, continued stirred reactors, etc. Photobioreactors in general provide better control of cultivation conditions, yield higher productivity and reproducibility, reduce contamination risk, and allow greater selection of algal species used for cultivation. The bioreactor has a photolimited central dark zone and a better lit peripheral zone close to the surface (Chisti 2007). CO_2-enriched air is sparged into the reactor creating a turbulent flow. Turbulent flow simultaneously circulates cells between the light and dark zones and assists the mass transfer of carbon dioxide and oxygen gases. The frequency of light and dark zone cycling is

4.2 Production Systems

depended on the intensity of turbulence, cell concentration, optical properties of the culture, tube diameter, and the external irradiance level (Chisti 2007). Regulation of carbon dioxide and dissolved oxygen levels in the bioreactor is another key element to algal growth. The highest cost for closed systems is the energy cost associated with the mixing mechanism (Wijffels 2008).

Tubular photobioreactors consist of transparent tubes that are made of flexible plastic or glass. Tubes can be arranged vertically, horizontally, inclined, helically, or in a horizontal thin-panel design. Tubes are generally placed in parallel to each other or flat above the ground to maximize the illumination surface-to-volume ratio of the reactor. The diameter of tubes is usually small and limited (0.2-m diameter or less) to allow light penetration to the center of the tube where the light coefficient and linear growth rate of culture decrease with increasing unit diameter (Ogbonna and Tanaka 1997; Riesing 2006). The growth medium circulates from a reservoir to the reactor and back to the reservoir. A turbulent flow is maintained in the reactor to ensure distribution of nutrients, improve gas exchange, minimize cell sedimentation, and circulate biomass for equal illumination between the light and dark zones.

The most widely used photobioreactor is a tubular design, which has a number of clear transparent tubes, usually aligned with the sun's rays. The tubes are generally less than 10 cm in diameter to maximize sunlight penetration. The medium broth is circulated through a pump to the tubes, where it is exposed to light for photosynthesis, and then back to a reservoir. A portion of the algae is usually harvested after it passes through the solar collection tubes, making continuous algal culture possible. In some photobioreactors, the tubes are coiled spirals to form what is known as a helical-tubular photobioreactor. The microalgal broth is circulated from a reservoir to the solar collector and back to the reservoir (Chisti 2007).

Figure 4.1 depicts a tubular photobioreactor with parallel run horizontal tubes (Campbell 2008). A tubular photobioreactor consists of an array of straight transparent tubes that are usually made of plastic or glass. This tubular array, or the solar

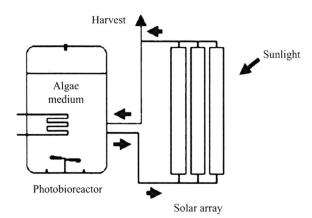

Figure 4.1 A tubular photobioreactor with parallel run horizontal tubes

collector, is where the sunlight is captured as seen in Figure 4.1. The solar collector tubes are generally 0.1 m or less in diameter. Tube diameter is limited because light does not penetrate too deeply in the dense culture broth, which is necessary for ensuring a high biomass productivity of the photobioreactor. Figure 4.2 shows another type of tubular photobioreactor with parallel run horizontal tubes (Chisti 2007). Figure 4.3 shows a tubular (a) and a vertical (b) photobioreactor.

Flat-plated photobioreactors are usually made of transparent material. The large illumination surface area allows high photosynthetic efficiency, low accumulation of dissolved oxygen concentration, and immobilization of algae (Ugwu *et al.* 2008). The reactors are inexpensive and easy to construct and maintain. However, the large surface area presents scale-up problems, including difficulties in controlling culture temperature and carbon dioxide diffusion rate and the tendency for algae adhering to the walls.

An inclined triangular tubular photobioreactor was designed to install adjacent to a power plant utilizing flue gas as the feed gas. Flue gas entered the reactor from the bottom of the inclined tube. Gas bubbles traveled along the inner surface of the

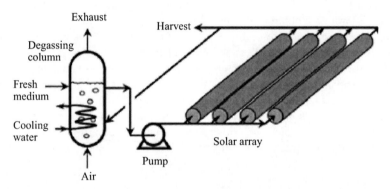

Figure 4.2 A tubular photobioreactor with parallel run horizontal tubes

Figure 4.3 (a) Tubular and (b) vertical photobioreactors

tube generating eddies for mixing and preventing fouling. The upper surface of the inclined tube absorbed natural light. The mixing to the algal culture and the flow rate of flue gas influences the growth rate of algae. The system worked, and 15 to 30% of algae were harvested each day. The setup was able to remove 82% of the carbon dioxide on a sunny day and 50% of the carbon dioxide on a cloudy day. Nitrogen oxide was also lowered by 86% (Riesing 2006).

Rectangular tanks are another example of photobioreactors. Unlike the circular tank design, rectangular tanks do not require a stirring device when a sufficiently high gas velocity is used. Drain pipes and gas spargers are located at the bottom of the tank.

Continuous stirred tank reactors (CSTRs) consist of a wide, hollow, capped cylindrical pipe that operates both indoors and outdoors with low contamination risk. A mechanical stirrer and a light source are inserted from the top of the reactor. Drain channels and gas injectors are positioned at the bottom (and midsection) of the reactor. The uniform turbulent flow established within the reactor promotes algal growth and prevents fouling.

Another photobioreactor uses helical coils made of plastic tubing placed across a columnlike structure. A group of helical coils make up one unit of photobioreactor. Each helical coil runs independently with its own gas injector, pump, and gas removal system. The helical coils operate both indoors (fluorescent light) and outdoors (sunlight).

Similar to the helical coils, square tubular reactors consist of plastic tubing arranged in a series of squares. One pump is used to provide algal flow through the series of squares and back. Compared with the helical coils, the square tubing is longer and holds more algal volume. The unit is also intended to be installed on the rooftop of a power plant and can operate both outdoor and indoor. However, light is a limitation where only one side of the square is exposed to the Sun at a time. To maximize light penetration, square tubular reactors cannot be packed as closely as the inclined triangular tubular reactor or the helical coils.

4.2.3 Open-pond Systems

Open ponds are the oldest and simplest systems for mass cultivation of microalgae. The pond is designed in a raceway configuration, in which a paddlewheel circulates and mixes the algal cells and nutrients. The raceways are typically made from poured concrete, or they are simply dug into the earth and lined with a plastic liner to prevent the ground from soaking up the liquid. Baffles in the channel guide the flow around the bends in order to minimize space. The system is often operated in a continuous mode, i.e., the fresh feed is added in front of the paddlewheel, and the algal broth is harvested behind the paddlewheel after it has circulated through the loop.

Open-pond systems are shallow ponds in which algae are cultivated. Nutrients can be provided through runoff water from nearby land areas or by channeling the water from sewage/water treatment plants (Carlsson et al. 2007). The water is typically kept in motion by paddlewheels or rotating structures, and some mixing can be accomplished by appropriately designed guides. Algal cultures can be defined (one or more selected strains), or are made up of an undefined mixture of strains. For an overview of systems used see Borowitzka (1999) and Chaumont (1993).

The only practicable methods of large-scale production of microalgae are raceway ponds (Terry and Raymond 1985; Molina Grima 1999) and tubular photobioreactors (Molina Grima et al. 1999; Tredici 1999; Sánchez Mirón et al. 1999).

Open architecture approaches (e.g., ponds or traditional racetracks), while possibly the cheapest of all current techniques, suffer challenges with contamination, evaporation, temperature control, CO_2 utilization, and maintainability.

Figure 4.4 shows open-pond "algae farm" systems. The ponds are "raceway" designs in which the algae, water, and nutrients circulate around a racetrack. Paddlewheels provide the flow. The algae are thus kept suspended in water. Algae are circulated back up to the surface on a regular frequency. The ponds are kept shallow because of the need to keep the algae exposed to sunlight and the limited depth to which sunlight can penetrate the pond water. The ponds are operated continuously; that is, water and nutrients are constantly fed to the pond while algae-containing water is removed at the other end. Some kind of harvesting system is required to recover the algae, which contain substantial amounts of natural oil.

Large-scale outdoor culture of microalgae and cyanobacteria in open ponds, raceways, and lagoons is well established (Becker 1994). Open culture is used commercially in the USA, Japan, Australia, India, Thailand, China, Israel, and elsewhere to produce algae for food, feed, and extraction of metabolites. Open-culture systems allow relatively inexpensive production but are subject to contamination. Consequently, only a few algal species can be cultured in open outdoor systems. Species

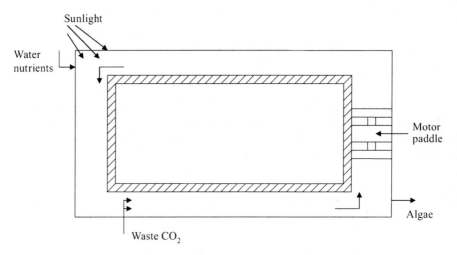

Figure 4.4 Open-pond "algae farm" systems

4.2 Production Systems

that grow successfully in the open include rapid growers such as *Chlorella* and species that require a highly selective extremophilic environment that does not favor the growth of most potential contaminants. For example, species such as *Spirulina* and *Dunaliella* thrive in highly alkaline and saline selective environments, respectively. Algae produced in quantities in open systems include *Spirulina, Chlorella, Dunaliella, Haematococcus, Anabaena*, and *Nostoc* (Chisti 2006).

Algae farms are large ponds. The size of these ponds is measured in terms of surface area, since surface area is so critical to capturing sunlight. Their productivity is measured in terms of biomass produced per day per unit of available surface area. Even at levels of productivity that would stretch the limits of an aggressive research and development program, such systems require acres of land. At such large sizes, it is more appropriate to think of these operations on the scale of a farm. Such algae farms would be based on the use of open, shallow ponds in which some source of waste CO_2 could be efficiently bubbled into the ponds and captured by the algae. Careful control of pH and other physical conditions for introducing CO_2 into the ponds allows for more than 90% utilization of injected CO_2. Raceway ponds, usually lined with plastic or cement, are about 15 to 35 cm deep to ensure adequate exposure to sunlight. They are typically mixed with paddlewheels, are usually lined with plastic or cement, and are between 0.2 to 0.5 ha in size. Paddlewheels provide motive force and keep the algae suspended in the water. The ponds are supplied with water and nutrients, and mature algae are continuously removed at one end.

Figure 4.5 shows a raceway pond (Chisti 2007). A raceway pond is made up of a closed-loop recirculation channel that is typically about 0.3 m deep. As shown in the figure, mixing and circulation are produced by a paddlewheel. Flow is guided

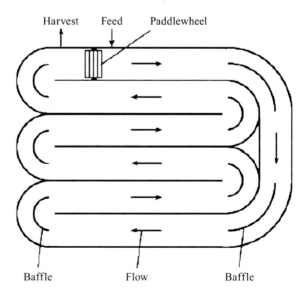

Figure 4.5 Arial view of a raceway pond

around bends by baffles placed in the flow channel, and raceway channels are built in concrete or compacted earth and may be lined with white plastic. During daylight, the culture is fed continuously in front of the paddlewheel where the flow begins. Broth is harvested behind the paddlewheel on completion of the circulation loop. The paddlewheel operates all the time to prevent sedimentation.

Photosynthesis is the most important biochemical process in which plants, algae, and some bacteria harness the energy of sunlight to produce food. Organisms that produce energy through photosynthesis are called photoautotrophs. Photosynthesis is a process in which green plants utilize the energy of sunlight to manufacture carbohydrates from carbon dioxide and water in the presence of chlorophyll (Viswanathan 2006).

Raceway ponds for mass culture of microalgae have been used since the 1950s. Extensive experience exists on the operation and engineering of raceways. The largest raceway-based biomass production facility occupies an area of 440,000 m^2 (Spolaore *et al.* 2006). Productivity is affected by contamination with unwanted algae and microorganisms that feed on algae. Raceway ponds and other open culture systems for producing microalgae are further discussed by Terry and Raymond (1985).

There are quite a number of sources of waste CO_2. Every operation that involves the combustion of fuel for energy is a potential source. One program targeted coal and other fossil-fuel-fired power plants as the main sources of CO_2. Typical coal-fired power plants emit flue gas from their stacks containing up to 13% CO_2. This high concentration of CO_2 enhances transfer and uptake of CO_2 in ponds. Figure 4.6 shows the production of algae in open ponds.

For large-scale biofuel production, which would require systems of hundreds of hectares in scale, this would mean deploying tens of thousands of such repeating units, at great capital and operating cost. Open ponds, specifically mixed raceway ponds, are much cheaper to build and operate, can be scaled up to several hectares for individual ponds, and are the method of choice for commercial microalgae production. However, such open ponds also suffer from various limitations, including

Figure 4.6 Production of algae in open ponds

4.2 Production Systems

more rapid (than closed systems) biological invasions by other algae, algae grazers, fungi and amoeba, etc., and temperature limitations in cold or hot humid climates.

Microalgae can be cultivated in coastal areas. The raceway pond system of biomass culture must be approved to achieve high and sustained growth rates and oil yields that are essential to developing an algal-based biofuel industry.

4.2.4 Closed and Hybrid Systems

Other system designs for algae production are possible. The Japanese, French, and German governments have invested significant R&D dollars on novel closed bioreactor designs for algae production. The main advantage of such closed systems is that they are not as subject to contamination with whatever organism happens to be carried in the wind.

When designing a photobioreactor, design parameters such as reactor dimension, flowrate, light requirements, culture condition, algae species, reproducibility, and economic value need to be taken into consideration. Depending on the reactor dimensions, site location, and local climate, these parameters can determine the type of cultivation system needed (open versus closed). Reactor design should have good mixing properties, efficiency, and reproducibility and be easy to maintain and sterilize. An efficient photobioreactor not only improves productivity but also is used to cultivate multiple strains of algae. The performance of a photobioreactor is measured by volumetric productivity, areal productivity, and productivity per unit of illuminated surface (Riesing 2006). Volumetric productivity is a function of biomass concentration per unit volume of bioreactor per unit of time. Areal productivity is defined as biomass concentration per unit of occupied land per unit of time. Productivity per unit of illuminated surface is measured as biomass concentration per area per unit of time.

Closed bioreactors support up to fivefold higher productivity with respect to reactor volume and consequently have a smaller "footprint" on a yield basis. Besides saving water, energy, and chemicals, closed bioreactors have many other advantages that are increasingly making them the reactor of choice for biofuel production, as their costs are lower (Schenk *et al.* 2008). Closed bioreactors permit essentially single-species culture of microalgae for prolonged periods. Most closed bioreactors are designed as tubular reactors, plate reactors, or bubble column reactors (Weissman *et al.* 1988; Pulz 2001). Other less common designs like semihollow spheres have been reported to run successfully (Sato *et al.* 2006).

Enclosed photobioreactors have been employed to overcome the contamination and evaporation problems encountered in open ponds (Molina Grima *et al.* 1999). These systems are made of transparent materials and are generally placed outdoors for illumination by natural light. The cultivation vessels have a large surface-area-to-volume ratio.

The main problems in the large-scale cultivation of microalgae outdoors in open ponds are low productivity and contamination. To overcome these problems, a closed system consisting of polyethylenes sleeves was developed. In a study conducted outdoors. The closed system was found to be superior to open ponds with respect to growth and production in a number of microalgae. In both closed and open systems, growth and production under continuous operation were higher than in batch cultivation (Cohen *et al.* 1991).

Sananurak *et al.* (2009) have designed, built, and operated a closed, recirculating, continuous culture system to produce microalgae and rotifers in seawater (25% salinity) for larval fish culture. The system opens up a new perspective in terms of automated production of rotifers without labor cost. Rotifers can be easily harvested daily by a conical harvest net, and there is no routine maintenance work. This new, automated system has three components: a microalgae culture, a rotifer culture and storage with harvest, and a water treatment and reuse component.

The preferred alternative is closed photobioreactors, where the algae fluid remains in a closed environment to enable accelerated growth and better control over environmental conditions. These glass or plastic enclosures, often operated under modest pressure, can be mounted in a variety of horizontal or vertical configurations and can take many different shapes and sizes. Rigid frameworks or structures are usually used to support the photobioreactor enclosures.

Open systems using a monoculture are also vulnerable to viral infection. The energy that a high-oil strain invests in the production of oil is energy that is not invested in the production of proteins or carbohydrates, usually resulting in the species being less hardy or having a slower growth rate. Algal species with lower oil content, not having to divert their energies away from growth, have an easier time in the harsher conditions of an open system.

Closed systems (not exposed to open air) do not have the problem of contamination by other organisms blown in by the air. The problem for a closed system is finding a cheap source of sterile CO_2. Several experimenters have found the CO_2 from a smokestack works well for growing algae.

In hybrid systems, both open ponds as well as closed bioreactor system are used in combination to get better results. Open ponds are a very proficient and lucrative method of cultivating algae, but they become contaminated with superfluous species very quickly. A combination of both systems is probably the most logical choice for cost-effective cultivation of high yielding strains for biofuels. Open ponds are inoculated with a desired strain that had invariably been cultivated in a bioreactor, whether it is as simple as a plastic bag or a high-tech fiber-optic bioreactor. Importantly, the size of the inoculums needs to be large enough for the desired species to establish in the open system before an unwanted species. Therefore, to minimize contamination issues, cleaning or flushing the ponds should be part of the aquaculture routine, and as such, open ponds can be considered as batch cultures (Schenk *et al.* 2008).

Abundant light, which is necessary for photosynthesis, is the third requirement. This is often accomplished by situating the facility in a geographic location with abundant, uninterrupted sunshine such as the American Southwest (Brown and Zeiler 1993). This is a favored approach when cultivating in open ponds. When working with bioreactors, sunlight quantity and quality can be further enhanced through the use of solar collectors, solar concentrators, and fiber optics in a system called photobioreactors (Scott and Bryner 2006; Chisti 2007). These technologies allow optimal sunlight to reach algal cells either by allowing them to float in arrays of thin, horizontal tubes or by directing light, through a fiber-optic matrix, through the bioreactor chamber itself.

The pH level generally increases as the microalgae consume CO_2. Addition of carbon dioxide along the reactor would sustain microalgal growth by preventing carbon limitation and an excess rise in pH. However, tubular photobioreactors do not work well in large-scale production because the surface-to-volume ratio is lower, causing poor light absorption. Length of tubes is another concern of tubular photobioreactors. As the tube length increases, the time for microalgae exposure to light increases, hence increasing the absorption of available CO_2 and increasing photosynthesis rate. However, the dissolved oxygen level also increases, which can easily lead to oxygen poisoning, and photoinhibition can result from the excess light exposure (Ogbonna and Tanaka 1997).

4.3 Production Costs

Producing microalgal biomass is generally more expensive than growing crops. Photosynthetic growth requires light, carbon dioxide, water, and inorganic salts. Temperature must remain generally within a range of 293 to 303 K. To minimize expense, biodiesel production must rely on freely available sunlight, despite daily and seasonal variations in light levels (Chisti 2007).

Algae cultivation has four basic, and equally important, requirements: carbon, water, light, and space. By maximizing the quality and quantity of these requirements, it is possible to maximize the quantity of oil-rich biomass and the return on investment. In order to maximize algal growth, CO_2 needs to be provided at very high levels, much higher than can be attained under natural conditions. Rather than becoming an expense, this need for CO_2 fertilization creates a unique opportunity to offset costs by consuming air pollution. The flue gases from industrial processes, and in particular from power plants, are rich in CO_2 that would normally be released directly into the atmosphere and thereby contribute to global warming. By diverting the CO_2 fraction of the flue gas through an algae cultivation facility, the CO_2 can be diverted back into the energy stream and the rate of algal production can be greatly increased (Pulz 2007). Although most of the CO_2 will ultimately be deposited in the atmosphere, we can realize a greater energy return for each molecule of carbon.

Water, containing the essential salts and minerals for growth, is the second requirement. Fresh water is a valuable resource as are the salts and minerals needed; however, algae cultivation can be coupled to another type of environmental remediation that will enhance productivity while mitigating pollution. High nutrient wastewater from domestic or industrial sources, which may already contain nitrogen and phosphate salts, can be added to the algal growth medium directly (Schneider 2006).

This allows for inexpensive improvement in algae production along with simultaneous treatment of wastewater. Alternatively, salt water can be used, either from a saline aquifer or sea water. This means that competition for water will be low.

The main advantages of second-generation microalgal systems are that they (1) have a higher photon conversion efficiency, (2) can be harvested batchwise nearly year round providing a reliable and continuous supply of oil, (3) can utilize salt and wastewater streams, thereby greatly reducing freshwater use, (4) can couple CO_2-neutral fuel production with CO_2 sequestration, and (5) produce nontoxic and highly biodegradable biofuels. Current limitations exist mainly in the harvesting process and in the supply of CO_2 for high efficiency production (Schenk *et al.* 2008).

Prior economic-engineering feasibility analyses have concluded that even the simplest open-pond systems, including harvesting and algal biomass processing equipment, would cost at least \$ 100,000 per hectare, and possibly significantly more. To this would need to be added operating costs. Algae production requires a site with favorable climate, available water (which can be saline, brackish, or wastewater), a ready and essentially free source of CO_2, nearly flat land, and a clay soil or liner, as plastic liners would be too expensive.

Nutrients such as phosphorus must be supplied in significant excess from phosphates complex with metal ions; therefore, not all the added phosphorus is bioavailable. Sea water supplemented with commercial nitrate and phosphate fertilizers and a few other micronutrients is commonly used for growing marine microalgae (Molina Grima *et al.* 1999). Genetic and metabolic engineering is likely to have the greatest impact on improving the economics of production of microalgal diesel (Roessler *et al.* 1994; Dunahay *et al.* 1996).

Growth media are generally inexpensive. Microalgal biomass contains approx. 50% carbon by dry weight (Sánchez Mirón *et al.* 2003). All of this carbon is typically derived from CO_2. Producing 100 tons of algal biomass fixes roughly 183 tons of carbon dioxide. Feeding controlled in response to signals from pH sensors minimizes loss of CO_2 and pH variations.

Algae can grow practically anywhere where there is enough sunlight. Some algae can grow in saline water. All algae contain proteins, carbohydrates, lipids, and nucleic acids in varying proportions. While the percentages vary with the type of algae, there are algae types whose overall mass is comprised of up to 40% fatty acids (Becker 1994). The most significant distinguishing characteristic of algal oil is its yield and, hence, its biodiesel yield. According to some estimates, the yield (per acre) of oil from algae is over 200 times the yield from the best-performing plant/vegetable oils (Sheehan *et al.* 1998). Microalgae are the fastest-growing photosynthesizing organisms. They can complete an entire growing cycle every few days. Approximately 46 tons of oil/ha/year can be produced from diatom algae. Different algae species produce different amounts of oil. Some algae produce up to 50% oil by weight. The production of algae to harvest oil for biodiesel has not been undertaken on a commercial scale, but working feasibility studies have been conducted to arrive at the above number. Specially bred mustard varieties can produce reasonably high oil yields and have the added benefit that the meal left over

4.3 Production Costs

after the oil has been pressed out can act as an effective and biodegradable pesticide (Demirbas 2009c).

Microalgae are very efficient solar energy converters and can produce a great variety of metabolites (Chaumont 1993). A culture of algae can yield 30 to 50% oil. Oil supply is based on the theoretical claims that 47,000 to 308,000 L/ha/year of oil could be produced using algae. The calculated cost per barrel would be only $ 20. Currently, a barrel of oil in the US market is selling for over $ 100 per barrel. Despite all the claims and research dating from the early 1970s, none of the projected algae and oil yields has been achieved (Dimitrov 2008; Demirbas 2009c). Algae, like all plants, require large quantities of nitrogen fertilizer and water, plus significant fossil energy inputs for a functioning system (Goldman and Ryther 1977).

Harvesting algae from tanks and separating the oil from the algae is a difficult and energy-intensive process (Pimentel et al. 2004). One difficulty in culturing algae is that the algae shade one another and thus there are different levels of light saturation in the cultures, even under Florida conditions. This influences the rate of growth of the algae. In addition, wild strains of algae invade and dominate algae culture strains, and oil production by the algae is reduced (Biopact 2008). Another major problem with the culture of algae in ponds or tanks is the harvesting of the algae. This problem was observed at the University of Florida where algae were being cultured in managed ponds for the production of nutrients for hogs. After 2 years without success, the algal-nutrient culture was abandoned (Pimentel 2008).

In recent years, there has been increasing interest in greenhouse gas mitigation technologies. As a consequence, there has been renewed interest in microalgae mass culture and fuel production from the perspective of CO_2 utilization. This is not a new concept, as Oswald and Golueke (1960) had previously emphasized the potential for microalgae systems to reduce and avoid CO_2 emissions and thus reduce the potential for global warming. Indeed, microalgae have a rather unique attribute: they can utilize concentrated CO_2 for growth, rather than the atmospheric levels of CO_2 used by higher plants. Flue gas can be utilized in algal ponds.

Microalgae wastewater treatment uses less energy, and thus fossil fuels, than conventional treatment processes, resulting in a reduction in greenhouse gas emissions. Wastewater treatment processes could provide a near-term pathway to developing large-scale microalgae production processes and could find applications in the real world.

Open ponds can be categorized into natural waters (lakes, lagoons, ponds) and artificial ponds or containers. The most commonly used systems include shallow big ponds, tanks, circular ponds, and raceway ponds. The major capital costs for an open-pond system are tabulated in Table 4.1 (Weissman and Goebel 1987; Sheehan et al. 1998). Polymers can be used in very small amounts, without contributing a major cost to the overall process. The base case ($30 g/m^2/d$) capital costs were estimated at almost $ 72,000/ha, without working capital, or almost twice as high as the prior effort (Benemann et al. 1982). This was due to higher costs for many components, such as earthworks, which were several-fold higher. Among other things, higher costs were assumed for rough and fine (laser) grading, which depends on the type of site available. Also, the 1987 study estimated about $ 5,000/ha to provide

Table 4.1 Capital costs for an open-pond system

Capital costs ($/ha)	112 mt/ha/year	224 mt/ha/year
Growth bonds		
Earthworks	10,135	10,135
Wall and structural	8,304	8,304
Carbonation system	1,830	2,978
Instrumentation	500	500
Primary (settling ponds)	7,479	7,479
Secondary (centrifuges)	3,958	6,467
Systemwide costs		
Water supply/distribution	4,426	4,426
CO_2 distribution	260	421
Nutrient supply	781	781
Salt disposal	833	833
Buildings	573	573
Roads and drainage	521	521
Electrical distribution/supply	1,924	2,215
Machinery	417	417
Eng. + contract. (25% above)	11,715	12,742
Land costs ($ 1,250/ha)	2,500	2,500
Gen-set (eng. + cont. included)	8,250	16,500
Anaerobic digestion	3,627	7,254
Total capital costs ($/ha)	72,952	89,965

Table 4.2 Operating costs for an open-pond system

Operating costs ($/ha/y)	112 mt/ha/year	224 mt/ha/year
CO_2 (2 kg/2 kg biomass)	6,290	12,580
P superphosphate, Fe as $FeSO_4$	530	1,070
Flocculants	1,120	2,250
Power mixing (10,730 kWh/ha)	700	700
1E harvest (1,770 kWh/ha)	120	120
2E harvest (5,730 kWh/ha)	370	600
Water supply (8,750 kWh/ha)	570	570
Other (1,562 kWh/ha)	110	110
Power production (6.5 c/kWh)	(2,250)	(5,100)
Salt disposal ($ 67/mt)	1,130	1,130
Maintenance	1,970	2,940
Labor	1,390	1,390
Total operating cost	12,420	19,110
Capital cost (25%/y of total)	18,238	22,491
Total cost ($/ha/y)	30,658	41,601
Total biomass cost ($/mt)	270	185

a 3- to 5-cm crushed-rock layer to reduce the suspension of silt from the pond bottom. There is, however, little evidence that such erosion prevention is needed, except perhaps for some areas around the paddlewheel and perhaps the turns. Further, the

Table 4.3 Comparative economics of open ponds and closed photobioreactors (PBRs)

Parameter	Relative advantage
Capital/operating costs	Open ponds $\ll$ PBRs
Biomass concentration	Open ponds < PBRs
Oxygen inhibition	Open ponds > PBRs
Contamination risk	Open ponds < PBRs
Water losses	Open ponds $\sim$ PBRs
Carbon dioxide losses	Open ponds $\sim$ PBRs
Process control	Open ponds $\sim$ PBRs
Space required	Open ponds $\sim$ PBRs

Weissman and Goebel (1987) study selected slip form poured concrete walls and dividers (baffles) as the design of choice. A power generation system can be specified to produce electricity from the methane generated from the algal residues (at about 10% of total costs).

Table 4.2 shows the operating costs for an open-pond system (Sheehan *et al.* 1998). The operating costs were discussed in terms of mixing, carbon utilization, nutrient, flocculants, salt disposal, maintenance, labor, and the accumulation of photosynthetically produced oxygen (Benemann 2008).

Table 4.3 shows the comparative economics of open ponds and closed photobioreactors.

References

Becker, E. W. 1994. Photobioreactors: production systems for photo-trophic microorganisms. In: Baddiley, J., Carey, N. H., Higgins, I. J., Potter, W. G. (eds.). Microalga: Biotechnology and Microbiology. Cambridge University Press, Cambridge, UK.

Benemann, J. R., Augenstein, D. C., Weissman, J. C. 1982. Microalgae as a source of liquid fuels, appendix: technical feasibility analysis. Final Report to the US Department of Energy.

Benemann, J. R. 2008. Open ponds and closed photobioreactors – comparative economics. 5th Annual World Congress on Industrial Biotechnology and Bioprocessing. Chicago, 30 April 2008.

Biopact. 2008. An in-depth look at biofuels from algae. Available at http://news.mongabay.com/bioenergy/2007/01/in-depth-look-at-biofuels-from-algae.html. Last accessed 18 December 2009.

Borowitzka, M. A. 1999. Commercial production of microalgae: ponds, tanks, tubes and fermenters. J Biotechnol 70:313–321.

Bosma, R., van Spronsen, W. A., Tramper, J., Wijffels, R. H. 2003. Ultrasound, a new separation technique to harvest microalgae. J Appl Phycol 15:143–153.

Brown, L. M., Zeiler, B. G. 1993. Aquatic biomass and carbon dioxide trapping. Energy Convers Manage 34:1005–1013.

Campbell, M. N. 2008. Biodiesel: algae as a renewable source for liquid fuel. Guelph Eng J 1:2–7.

Carlsson, A. S., van Beilen, J. B., Möller, R., Clayton, D. 2007. Micro- and macro-algae: utility for industrial applications. In: Bowles, D. (ed.). Outputs from the EPOBIO: Realising the Economic Potential of Sustainable Resources – Bioproducts from Non-food Crops Project, CNAP, University of York, UK.

Carvalho, A. P., Meireles, L. A., Malcata, F. X. 2006. Microalgal reactors: a review of enclosed system designs and performances. Biotechnol Prog 22:1490–1506.

Chaumont, D. 1993. Biotechnology of algal biomass production: a review of systems for outdoor mass culture. J Appl Phycol 5:593–604.

Chisti, Y. 2006. Microalgae as sustainable cell factories. Environ Eng Manage J 5:261–274.

Chisti, Y. 2007. Biodiesel from microalgae. Biotechnol Adv 25:294–306.

Choi, S. L., Suh, I. S., Lee, C. G. 2003. Lumostatic operation of bubble column photobioreactors for Haematococcus pluvialis cultures using a specific light uptake rate as a control parameter. Enzyme Microbial Technol 33:403–409.

Cohen, E., Koren, A., Arad, S. M. 1991. A closed system for outdoor cultivation of microalgae. Biomass Bioenergy 1:83–88.

Csordas, A., Wang, J. K. 2004. An integrated photobioreactor and foam fractionation unit for the growth and harvest of *Chaetoceros* spp. in open systems. Aquacult Eng 30:15–30.

Demirbas, A. 2006. Oily products from mosses and algae via pyrolysis. Energy Sources A 28:933–940.

Demirbas, A. H. 2009a. Inexpensive oil and fats feedstocks for production of biodiesel. Energy Educ Sci Technol A 23:1–13.

Demirbas, A. 2009b. Production of biodiesel from algae oils. Energy Sources A 31:163–168.

Demirbas, A. H. 2009c. Inexpensive oil and fats feedstocks for production of biodiesel. Energy Educ Sci Technol A 23:1–13.

Dimitrov, K. 2008. Green fuel technologies: a case study for industrial photosynthetic energy capture. Brisbane, Australia. Available at http://www.nanostring.net/Algae/.

Dunahay, T. G., Jarvis, E. E., Dais, S. S., Roessler, P. G. 1996. Manipulation of microalgal lipid production using genetic engineering. Appl Biochem Biotechnol 57–58:223–231.

Goldman, J. C., Ryther, J. H. 1977. Mass production of algae: bio-engineering aspects. In: Mitsui, A. *et al.* (eds.). Biological Solar Energy Conversion, Academic, New York.

Grobbelaar, J. U. 2004. Algal nutrition. In: Richmond, A. (ed.). Handbook of Microalgal Culture: Biotechnology and Applied Phycology. Blackwell, London.

Haesman, M., Diemar, J., O'Connor, W., Soushames, T., Foulkes, L. 2000. Development of extended shelf-life microalgae concentrate diets harvested by centrifugation for bivalve molluscs – a summary. Aquacult Res 31:637–659

Hankamer, B., Lehr, F., Rupprecht, J., Mussgnug, J. H., Posten, C., Kruse, O. 2007. Photosynthetic biomass and H production by green algae: from bioengineering to bioreactor scale up. Phys Plant 131:10–21.

Janssen, M., Tramper, J., Mur, L. R., Wijffels, R. H. 2003. Enclosed outdoor photobioreactors: light regime, photosynthetic efficiency, scale-up and future prospects. Biotechnol Bioeng 81:193–210.

Knuckey, R. M., Brown, M. R., Robert, R., Frampton, D. M. F. 2006. Production of microalgal concentrates by flocculation and their assessment as aquaculture feeds. Aquacult Eng 35:300–313.

Lee, Y. 2001. Microalgal mass culture systems and methods: their limitation and potential. J Appl Phycol 13:307–315.

McHugh, D. J. 2003. A guide to the seaweed industry. FAO Fisheries Technical Paper No. 441. FAO, Rome.

Molina Grima, E. 1999. Microalgae, mass culture methods. In: Flickinger, M. C., Drew, S. W. (eds.). Encyclopedia of Bioprocess Technology: Fermentation, Biocatalysis and Bioseparation, vol. 3. Wiley, New York.

Molina Grima, E., Acién Fernández, F. G., García Camacho, F., Chisti, Y. 1999. Photobioreactors: light regime, mass transfer, and scaleup. J Biotechnol 70:231–247.

Molina Grima, E. M., Belarbi, E. H., Fernandez, F. G. A., Medina, A. R., Chisti, Y. 2003. Recovery of microalgal biomass and metabolites: process options and economics. Biotechnol Adv 20:491–515.

Ogbonna, J. C., Tanaka, H. 1997. Industrial-size photobioreactors. Chemtech 27:43–49.

References

Oswald, W. J., Golueke, C. G. 1960. Biological transformation of solar energy. Adv Appl Microbiol 11:223–242.

Ozkurt, I. 2009. Qualifying of safflower and algae for energy. Energy Educ Sci Technol A 23:145–151.

Patil, V., Reitan, K. I., Knudsen, G., Mortensen, L., Kallqvist, T., Olsen, E., Vogt, G., Gislerød, H. R. 2005. Microalgae as source of polyunsaturated fatty acids for aquaculture. Curr Topics Plant Biol 6:57–65.

Patil, V., Tran, K.-Q., Giselrød, H. R. 2008. Towards sustainable production of biofuels from microalgae. Int J Mol Sci 9:1188–1195.

Pimentel, D., Berger, B., Filiberto, D., Newton, M., Wolfe, B., Karabinakis, B., Clark, S., Poon, E., Abbett, E., Nandagopal, S. 2004. Water resources: agricultural and enveronmental issues. Biosci 54: 909–918.

Pimentel, D. (ed.). 2008. Biofuels, solar and wind as renewable energy systems: benefits and risks. Springer, New York.

Pirt, S. J. 1986. The thermodynamic efficiency (quantum demand) and dynamics of photosynthetic growth. Appl Phycol 13:307–315.

Poelman, E., DePauw, N., Jeurissen, B. 1997. Potential of electrolytic flocculation for recovery of micro-algae. Resour Conserv Recyc 19:1–10.

Pulz, O. 2001. Photobioreactors: production systems for phototrophic microorganisms. Appl Microbiol Biotechnol 57:287–293.

Pulz, O. 2007. Evaluation of greenfuel's 3D matrix algal growth engineering scale unit: APS Redhawk Unit, Phoenix, AZ, IGV Institut für Getreideverarbeitung GmbH, June–July 2007.

Richmond, A. 2004. Principles for attaining maximal microalgal productivity in photobioreactors: an overview. Hydrobiologia 512:33–37.

Riesing, T. 2006. Cultivating algae for liquid fuel production. http://oakhavenpc.org/cultivating_algae.htm. Accessed February 2008.

Rossignol, N., Lebeau, T., Jaouen, P., Robert, J. M. 2000. Comparison of two membrane-photobioreactors, with free or immobilized cells, for the production of pigments by a marine diatom. Bioproc Eng 23:495–501.

Roessler, P. G., Brown, L. M., Dunahay, T. G., Heacox, D. A., Jarvis, E. E., Schneider, J. C., Talbot, S. G., Zeiler, K. G. 1994. Genetic-engineering approaches for enhanced production of biodiesel fuel from microalgae. ACS Symp Ser 566:255–270.

Sananurak, C., Lirdwitayaprasit, T., Menasveta, P. 2009. Development of a closed-recirculating, continuous culture system for microalga (*Tetraselmis suecica*) and rotifer (*Brachionus plicatilis*) production. Sci Asia 35:118–124.

Sánchez Mirón, A., Contreras Gómez, A., García Camacho, F., Molina Grima, E., Chisti, Y. 1999. Comparative evaluation of compact photobioreactors for large-scale monoculture of microalgae. J Biotechnol 70:249–270.

Sato, T., Usui, S., Tsuchiya, Y., Yutaka, K. 2006. Invention of outdoor closet type photobioreactor for microalgae. Energy Convers Manage 47:791–799.

Schneider, D. 2006. Grow your own? Would the widespread adoption of biomass-derived transportation fuels really help the environment? Am Sci 94:408–409.

Schenk, P. M., Thomas-Hall, S. R., Stephens, E., Marx, U., Mussgnug, J. H., Posten, C., Kruse, O., Hankamer, B. 2008. Second generation biofuels: high-efficiency microalgae for biodiesel production. Bioenergy Res 1:20–43.

Scott, A., Bryner, M. 2006. Alternative fuels: rolling out next-generation technologies. Chem Week December 20–27:17–21.

Sheehan, J., Dunahay, T., Benemann, J., Roessler, P. 1998. A look back at the U.S. Department of Energy's Aquatic Species Program – Biodiesel from Algae. National Renewable Energy Laboratory (NREL) Report: NREL/TP-580-24190. Golden, CO.

Spolaore, P., Joannis-Cassan, C., Duran, E., Isambert, A. 2006. Commercial applications of microalgae. J Biosci Bioeng 101:87–96.

Terry, K. L., Raymond, L. P. 1985. System design for the autotrophic production of microalgae. Enzyme Microb Technol 7:474–487.

Tredici, M. 1999. Bioreactors, photo. In: Flickinger, M. C., Drew, S. W. (eds.). Encyclopedia of Bioprocess Technology, Fermentation, Biocatalysis and Bioseparation. Wiley, New York.

Ugwu, C. U., Aoyagi, H., Uchiyama, H. 2008. Photobioreactors for mass cultivation of algae. Biores Technol 99:4021–4028.

Viswanathan, B. 2006. An introduction to energy sources. Indian Institute of Technology, Madras, India.

Wang, B., Li, Y., Wu, N., Lan, C. Q. 2008. CO_2 bio-mitigation using microalgae. Appl Microbiol Biotechnol 79:707–718.

Weissman, J. C., Goebel, R. P. 1987. Design and analysis of pond systems for the purpose of producing fuels. Report, Solar Energy Research Institute, SERI/STR-231-2840, Golden, CO.

Weissman, J., Goebel, R. P., Benemann, J. R. 1988. Photobioreactor design: mixing, carbon utilization, and oxygen accumulation. Biotechnol Bioeng 31:336–344.

Wijffels, R. H. 2008. Potential of sponges and microalgae for marine biotechnology. Trends Biotechnol 26:26–31.

Chapter 5
Energy from Algae

5.1 Introduction

The world has been confronted with an energy crisis due to the depletion of finite fossil fuel resources. The use of fossil fuels as energy is now widely accepted as unsustainable due to depleting resources and also due to the accumulation of greenhouse gases in the atmosphere.

Biomass provides a number of local environmental gains. Biomass resources include agricultural and forest residues, algae and grasses, animal manure, organic wastes, and biomaterials. Supply of these resources is dominated by traditional biomass used for cooking and heating, especially in rural areas of developing countries. Biomass mainly now represents only 3% of primary energy consumption in industrialized countries. However, much of the rural population in developing countries, which represents about 50% of the world's population, is reliant on biomass, mainly in the form of wood, for fuel.

Energy forestry crops have a much greater diversity of wildlife and flora than the alternative land use, which is arable or pasture land. In industrialized countries, it is expected that the main biomass processes utilized in the future will be direct combustion of residues and wastes for electricity generation, bioethanol and biodiesel as liquid fuels, and combined heat and power production from energy crops. The future of biomass electricity generation lies in biomass-integrated gasification/gas turbine technology, which offers high energy-conversion efficiencies. In the future, biomass will have the potential to provide a cost-effective and sustainable supply of energy, while at the same time aiding countries in meeting their greenhouse-gas-reduction targets. By the year 2050, it is estimated that 90% of the world's population will live in developing countries.

Prior to the establishment of the US Department of Energy's (DOE) Aquatic Species Program, very little work had been conducted on biofuel production from lipid-accumulating algae. While the general idea of using algae for energy production has been around for over 50 years (Meier 1955), the concept of using lipids derived from algal cells to produce liquid fuels arose more recently. The research of liquid fuel produced from microalgae was begun in the mid-1980s in 20 centuries

A. Demirbas, M. Fatih Demirbas, *Algae Energy*
DOI 10.1007/978-1-84996-050-2, © Springer 2010

(Xu *et al.* 2006). Aquatic biomass may represent a convenient solution because it has a higher growth rate than terrestrial plants. Microalgae have been extensively studied so far, as they can grow in both fresh- and saltwater environments. Algal biomass contains three main components: carbohydrates, proteins, and natural oils. Algae are a promising source of renewable energy.

Microalgae can potentially be employed for the production of biofuels in an economically effective and environmentally sustainable manner. Microalgae have been investigated for the production of a number of different biofuels including biodiesel, bio-oil, biosyngas, and biohydrogen. The production of these biofuels can be coupled with flue gas CO_2 mitigation, wastewater treatment, and the production of high-value chemicals. Developments in microalgal cultivation and downstream processing are expected to further enhance the cost effectiveness of biofuel from microalgae (Li *et al.* 2008).

Algae, like corn, soybeans, sugar cane, Jatropha, and other plants, use photosynthesis to convert solar energy into chemical energy. They store this energy in the form of oils, carbohydrates, and proteins. The plant oil can be converted into biodiesel; hence biodiesel is a form of solar energy. The more efficient a particular plant is at converting that solar energy into chemical energy, the better it is from a biodiesel perspective, and algae are among the most photosynthetically efficient plants on Earth.

Algae for biofuels have been studied for many years for the production of hydrogen, methane, vegetable oils (for biodiesel), hydrocarbons, and ethanol. Algal hydrogen production has been extensively researched for over three decades, but no mechanism for it has ever been demonstrated.

Algae can be used to produce biofuel, called algae fuel, algal fuel, or even third-generation biofuel. Compared with second-generation biofuels, algal fuels have a higher yield: they can produce 30 to 100 times more energy per hectare compared to terrestrial crops.

The advantages and disadvantages of biofuel production using microalgae are shown in Table 5.1. Among the advantages are that the high growth rate of microalgae makes it possible to satisfy massive demand on biofuels using limited land resources without causing potential biomass deficit, microalgal cultivation consumes less water than land crops, the tolerance of microalgae to high CO_2 content in gas streams allows high-efficiency CO_2 mitigation, microalgal farming could be potentially more cost effective than conventional farming, and nitrous oxide release could be minimized when microalgae are used for biofuel production.

Table 5.1 Advatages and disadvatages of biofuel production using microalgae

Advantages	Disadvantages
High growth rate	Low biomass concentration
Less water demand than land crops	Higher capital costs
High-efficiency CO_2 mitigation	
More cost-effective farming	
Minimization of nitrous oxide release	

On the other hand, one of the major disadvantages of microalgae for biofuel production is the low biomass concentration in a microalgal culture due to the limited light penetration, which in combination with the small size of algal cells makes the harvest of algal biomasses relatively costly. The higher capital costs and the rather intensive care required by a microalgal farming facility compared to a conventional agricultural farm is another factor that impedes the commercial implementation of the biofuels-from-microalgae strategy.

5.1.1 Historical Perspective

Historically, algae have been seen as a promising source of protein and have been actively cultured by humans for centuries, mainly for food. Growing algae as a source of protein on a large scale in open ponds was first conceived by German scientists during World War II (Soeder 1986). The first attempt in the USA to translate the biological requirements for algal growth into engineering specifications for a large-scale plant was made at the Stanford Research Institute (1948–1950). During 1951, Arthur D. Little made a further advance through the construction and operation of a Chlorella pilot plant for the Carnegie Institute (Burlew 1953). These studies eventually provided some of the most comprehensive early information on the growth, physiology, and biochemistry of algae.

Under certain growth conditions, many microalgae can produce lipids that are suitable for conversion into liquid transportation fuels. In the late 1940s, nitrogen limitation was reported to significantly influence microalgal lipid storage. Spoehr and Milner (1949) published detailed information on the effects of environmental conditions on algal composition and described the effect of varying nitrogen supply on the lipid and chlorophyll content of Chlorella and some diatoms. Investigations by Collyer and Fogg (1955) demonstrated that the fatty acid content of most green algae was between 10 and 30% DCW. Werner (1966) reported an increase in the cellular lipids of a diatom during silicon starvation. Coombs et al. (1967) reported that the lipid content of the diatom Navicula pelliculosa increased by about 60% during a 14-h silicon starvation period. In addition to nutrition, fatty acid and lipid composition and content were also found to be influenced by a number of other factors such as light (Constantopolous and Bloch 1967; Nichols 1965; Pohl and Wagner 1972; Rosenberg and Gouaux 1967) and low temperatures (Ackman et al. 1968). With the advent of the oil embargo in the early 1970s, a search for alternative energy sources set the stage for an almost 20-year research effort devoted to biofuel production from algal lipids.

5.2 Biofuels

The term biofuel refers to solid, liquid, or gaseous fuels that are predominantly produced from biorenewable or combustible renewable feedstocks. Liquid biofuels will be important in the future because they will replace petroleum fuels. The biggest difference between biofuels and petroleum feedstocks is oxygen content. Biofuels are nonpolluting, locally available, accessible, sustainable, and reliable fuels obtained from renewable sources.

There are two global biorenewable liquid transportation fuels that might replace gasoline and diesel fuel. These are bioethanol and biodiesel. Bioethanol is a good alternative fuel that is produced almost entirely from food crops. Biodiesel has become more attractive recently because of its environmental benefits.

Transport is one of the main energy consuming sectors. It is assumed that biodiesel is used as a fossil diesel replacement and that bioethanol is used as a gasoline replacement. Biomass-based energy sources for heat, electricity, and transportation fuels are potentially carbon dioxide neutral and recycle the same carbon atoms. Due to the widespread availability of biofuels, opportunities in biorenewable fuel technology can potentially employ more people than fossil-fuel-based technology.

Renewable liquid biofuels for transportation have recently attracted considerable attention in different countries around the world because of their renewability, sustainability, widespread availability, and biodegradability and the benefits they bring with respect to regional development, rural manufacturing jobs, and reduction in greenhouse gas emissions. Table 5.2 shows the major benefits of biofuels.

Biofuels can be classified based on their production technologies: first-generation biofuels (FGBs), second-generation biofuels (SGBs), third-generation biofuels (TGBs), and fourth-generation biofuels.

The FGBs refer to biofuels made from sugar, starch, vegetable oils, or animal fats using conventional technology. The basic feedstocks for the production of FGBs are often seeds or grains such as wheat, which yields starch that is fermented into

Table 5.2 Major benefits of biofuels

Economic impacts	Sustainability
	Fuel diversity
	Increased number of rural manufacturing jobs
	Increased income taxes
	Increased investments in plant and equipment
	Agricultural development
	International competitiveness
	Reduced dependency on imported petroleum
Environmental impacts	Greenhouse gas reductions
	Reduced air pollution
	Biodegradability
	Higher combustion efficiency
	Improved land and water use
	Carbon sequestration
Energy security	Domestic targets
	Supply reliability
	Reduced use of fossil fuels
	Ready availability
	Domestic distribution
	Renewability

5.2 Biofuels

bioethanol, or sunflower seeds, which are pressed to yield vegetable oil that can be used in biodiesel. Table 5.3 shows the classification of renewable biofuels based on their production technologies.

Table 5.3 Classification of renewable biofuels based on their production technologies

Generation	Feedstock	Example
First	Sugar, starch, vegetable oils, animal fats	Bioalcohols, vegetable oil, biodiesel, biosyngas, biogas
Second	Non food crops, wheat straw, corn, wood, solid waste, energy crop	Bioalcohols, bio-oil, bio-DMF, biohydrogen, bio-Fischer–Tropsch diesel, wood diesel
Third	Algae	Vegetable oil, biodiesel
Fourth	Vegetable oil, biodiesel	Biogasoline

SGBs and TGBs are also called advanced biofuels. SGBs are made from nonfood crops, wheat straw, corn, wood, and energy crops using advanced technology. Algae fuel, also called algal oil or TGB, is a biofuel from algae (Demirbas 2007). Algae are low-input/high-yield (30 times more energy per acre than land) feedstocks to produce biofuels using more advanced technology. On the other hand, an emerging fourth-generation is based on the conversion of vegetable oil and biodiesel into biogasoline using the most advanced technology.

There are some barriers to the development of biofuel production. They are technological, economical, supply, storage, safety, and policy barriers. Reducing these barriers is one of the driving factors in the government's involvement in biofuel and biofuel research and development. Production costs are uncertain and vary with the feedstock available. The production of biofuels from lignocellulosic feedstocks can be achieved through two very different processing routes: biochemical and thermochemical. There is no clear candidate for "best technology pathway" between the competing biochemical and thermochemical routes. Technical barriers for enzymatic hydrolysis include: low specific activity of current commercial enzymes, high cost of enzyme production, and lack of understanding of enzyme biochemistry and mechanistic fundamentals.

The major nontechnical barriers are restrictions or prior claims on use of land (food, energy, amenity use, housing, commerce, industry, leisure or designations as areas of natural beauty, special scientific interest, etc.), as well as the environmental and ecological effects of large areas of monoculture. For example, vegetable oils are a renewable and potentially inexhaustible source of energy with energy content close to that of diesel fuel. On the other hand, extensive use of vegetable oils may lead to other significant problems such as starvation in developing countries. The vegetable oil fuels were not acceptable because they were more expensive than petroleum fuels.

There are few technical barriers to building biomass-fired facilities at any scale, from domestic to around 50 MW, above which considerations of the availability

102   5 Energy from Algae

and cost of providing fuel become significant. In general, however, the capacity and generating efficiency of biomass plants are considerably less than those of modern natural-gas-fired turbine systems. The main nontechnical limitations to investment in larger systems are economic, or in some countries reflect planning conditions and public opinion, where a clear distinction may not be made between modern effective biomass energy plants and older polluting incinerator designs.

5.2.1 Biodiesel

Vegetable oil (m)ethyl esters, commonly referred to as biodiesel, are prominent candidates as alternative diesel fuels. The name biodiesel has been given to transesterified vegetable oil to describe its use as a diesel fuel (Demirbas 2002). There has been renewed interest in the use of vegetable oils for making biodiesel due to its less polluting and renewable nature as against the conventional diesel, which is a fossil fuel that will eventually be exhausted. Biodiesel is technically competitive with or offers technical advantages over conventional petroleum diesel fuel. The vegetable oils can be converted into their (m)ethyl esters via a transesterification process in the presence of a catalyst. Methyl, ethyl, 2-propyl, and butyl esters were prepared from vegetable oils through transesterification using potassium or sodium alkoxides as catalysts. The purpose of the transesterification process is to lower the viscosity of the oil. Ideally, transesterification is potentially a less expensive way of transforming the large, branched molecular structure of bio-oils into smaller, straight-chain molecules of the type required in regular diesel combustion engines.

Biodiesel is a domestic fuel for diesel engines derived from natural oils like soybean oil. It is the name given to a variety of ester-based oxygenated fuel from renewable biological sources that can be made from processed organic oils and fats. The inedible oils such as jatropha curcas, madhuca indica, ficus elastica, azardirachta indica, calophyllum inophyllum jatropha, neem, pongamia pinnata, rubber seed, mahua, silk cotton tree, tall oil, microalgae etc. are easily available in developing countries and are very economical comparable to edible oils. Biodiesel obtained from waste cooking vegetable oils, tallow fat, and poultry fat have been considered promising options. Waste cooking oil is available at relatively cheap prices for biodiesel production in comparison with fresh vegetable oils.

The biodiesel esters were characterized for their physical and fuel properties including density, viscosity, iodine value, acid value, cloud point, pure point, gross heat of combustion, and volatility. The biodiesel fuels produced slightly lower power and torque and higher fuel consumption than No. 2 diesel fuel. Biodiesel is better than diesel fuel in terms of sulfur content, flash point, aromatic content, and biodegradability (Bala 2005).

Most of the biodiesel that is currently made uses soybean oil, methanol, and an alkaline catalyst. The high value of soybean oil as a food product makes production of a cost-effective fuel very challenging. However, there are large amounts of low-cost oils and fats such as restaurant waste and animal fats that could be converted into biodiesel. The problem with processing these low-cost oils and fats is that they often contain large amounts of free fatty acids (FFA) that cannot be converted into biodiesel using an alkaline catalyst (Demirbas 2003).

Biodiesel is an environmentally friendly alternative liquid fuel that can be used in any diesel engine without modification. There has been renewed interest in the use of vegetable oils for making biodiesel due to its less polluting and renewable nature as against the conventional petroleum diesel fuel. If biodiesel is used for engine fuel, this would in turn benefit the environment and local populations.

Microalgae contain oils, or lipids, that can be converted into biodiesel. The idea of using microalgae to produce fuel is not new, but has received renewed attention recently in the search for sustainable energy. Biodiesel is typically produced from plant oils, but there are widely voiced concerns about the sustainability of this practice. Biodiesel produced from microalgae is being investigated as an alternative to conventional crops, such as rapeseed: microalgae typically produce more oil, consume less space, and can be grown on land unsuitable for agriculture. However, many technical and environmental issues, such as land use and fertilizer input, still need to be researched and large-scale commercial production has still not been attained.

Using microalgae as a source of biofuels could mean that enormous cultures of algae will be grown for commercial production, which would require large quantities of fertilizers. While microalgae are estimated to be capable of producing 10 to 20 times more biodiesel than rapeseed, they need 55 to 111 times more nitrogen fertilizer: 8 to 16 tons/ha/year. Such quantities of nitrogen and phosphorus could damage the environment. Additionally, it could limit the economic viability of microalgae. Nitrogen and phosphorus found in algal waste, after the oils have been extracted, must therefore be recycled.

Microalgae contain lipids and fatty acids as membrane components, storage products, metabolites, and sources of energy. Algae present an exciting possibility as a feedstock for biodiesel, especially when you realize that oil was originally formed from algae.

Algal oil is converted into biodiesel through a transesterification process. Oil extracted from algae is mixed with alcohol and an acid or a base to produce the fatty acid methylesters that makes up biodiesel (Chisti 2007).

Many algae are exceedingly rich in oil, which can be converted to biodiesel. The oil content of some microalgae exceeds 80% of dry weight of algae biomass. The use of algae as energy crops has the potential, due to their easy adaptability to growth conditions, of growing either in fresh or marine waters and avoiding the use of land. Furthermore, two thirds of earth's surface is covered with water, thus algae would truly be renewable option of great potential for global energy needs. Figure 5.1 shows world production of biodiesel from 1980 to 2008.

Possessing approximately identical energy potential with mineral diesel fuel, the bio-diesel engine has a number of essential advantages:

- It is not toxic, contains practically no sulfur or carcinogenic benzene;
- Decays in natural conditions;

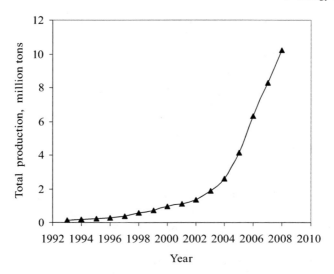

Figure 5.1 World production of biodiesel, 1980–2008

- Provides significant reduction in harmful emissions in the atmosphere upon burning, both in internal combustion engines and in technological units;
- Increases cetane number of fuel and its greasing ability, which essentially increases the engine performance;
- Has high ignition temperature (more than 373 K), which makes its use rather safe;
- It is derived from renewable resources.

5.2.2 Bioalcohol

Ethanol is the most widely used liquid biofuel. It is an alcohol and is fermented from sugars, starches, or cellulosic biomass. Most commercial production of ethanol is from sugar cane or sugar beet, as starches and cellulosic biomass usually require expensive pretreatment.

Carbohydrates (hemicelluloses and cellulose) in plant materials can be converted into sugars by hydrolysis. Fermentation is an anaerobic biological process in which sugars are converted into alcohol by the action of microorganisms, usually yeast. The resulting alcohol is ethanol. The value of any particular type of biomass as feedstock for fermentation depends on the ease with which it can be converted into sugars.

Bioethanol is a fuel derived from renewable sources of feedstock, typically plants such as wheat, sugar beet, corn, straw, and wood. Bioethanol is a petrol additive/substitute. It is possible that wood, straw, and even household wastes may be economically converted into bioethanol.

5.2 Biofuels

Bioethanol can be used as a 5% blend with petrol under EU quality standard EN 228. This blend requires no engine modification and is covered by vehicle warranties. With engine modification, bioethanol can be used at higher levels, for example, E85 (85% bioethanol). Figure 5.2 shows world ethanol production from 1980 to 2008 (RFA 2009).

Bioethanol can be produced from a large variety of carbohydrates with a general formula of $(CH_2O)_n$. Fermentation of sucrose is performed using commercial yeast such as *Saccharomyces ceveresiae*. The chemical reaction is composed of enzymatic hydrolysis of sucrose followed by fermentation of simple sugars. First, invertase enzyme in the yeast catalyzes the hydrolysis of sucrose to convert it into glucose and fructose. Second, zymase, another enzyme also present in yeast, converts the glucose and the fructose into ethanol.

Glucoamylase enzyme converts the starch into D-glucose. The enzymatic hydrolysis is then followed by fermentation, distillation, and dehydration to yield anhydrous bioethanol. Corn (60 to 70% starch) is the dominant feedstock in the starch-to-bioethanol industry worldwide.

Carbohydrates (hemicelluloses and cellulose) in lignocellulosic materials can be converted into bioethanol. The lignocellulose is subjected to delignification, steam explosion, and dilute acid prehydrolysis, which is followed by enzymatic hydrolysis and fermentation into bioethanol. A major processing step in an ethanol plant is enzymatic saccharification of cellulose to sugars through treatment by enzymes; this step requires lengthy processing and normally follows a short-term pretreatment step.

Hydrolysis breaks down the hydrogen bonds in the hemicellulose and cellulose fractions into their sugar components: pentoses and hexoses. These sugars can then

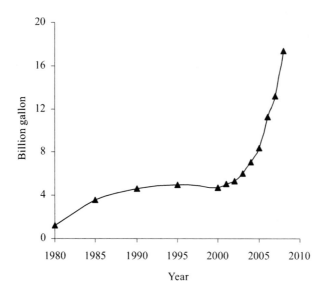

Figure 5.2 World ethanol production, 1980–2008

be fermented into bioethanol. The most commonly applied methods can be classified into two groups: chemical hydrolysis (dilute and concentrated acid hydrolysis) and enzymatic hydrolysis. In chemical hydrolysis, pretreatment and hydrolysis may be carried out in a single step. There are two basic types of acid hydrolysis processes commonly used: dilute acid and concentrated acid.

The biggest advantage of dilute acid processes is their fast rate of reaction, which facilitates continuous processing. Since 5-carbon sugars degrade more rapidly than 6-carbon sugars, one way to decrease sugar degradation is to have a two-stage process. The first stage is conducted under mild process conditions to recover the 5-carbon sugars, while the second stage is conducted under harsher conditions to recover the 6-carbon sugars.

Methanol, also known as "wood alcohol," is generally easier to find than ethanol. Sustainable methods of methanol production are currently not economically viable. Methanol is produced from synthetic gas or biogas and evaluated as a fuel for internal combustion engines. The production of methanol is a cost-intensive chemical process. Therefore, in current conditions, only waste biomass such as old wood or biowaste is used to produce methanol.

Before modern production technologies were developed in the 1920s, methanol was obtained from wood as a coproduct of charcoal production and, for this reason, was commonly known as wood alcohol. Methanol is currently manufactured worldwide by conversion or derived from syngas, natural gas, refinery off-gas, coal, or petroleum:

$$2H_2 + CO \rightarrow CH_3OH \qquad (5.1)$$

The chemical composition of syngas from coal and then from natural gas can be identical with the same H_2/CO ratio. A variety of catalysts are capable of causing the conversion, including reduced NiO-based preparations, reduced Cu/ZnO shift preparations, Cu/SiO_2 and Pd/SiO_2, and Pd/ZnO (Takezawa et al. 1987; Iwasa et al. 1993).

5.2.3 Costs, Prices, and Economic Impacts of Biofuels

In previous economic studies of biodiesel production, the main economic factors such as capital cost, plant capacity, process technology, raw material cost, and chemical costs were determined (Zhang et al. 2003). The major economic factor to consider for the input costs of biodiesel production is the feedstock, which is about 75 to 80% of the total operating cost. Other important costs are labor, methanol, and catalyst, which must be added to the feedstock (Demirbas 2003). Using an estimated process cost, exclusive of feedstock cost, of US$ 0.158/L for biodiesel production, and estimating a feedstock cost of US$ 0.539/L for refined soy oil, an overall cost of US$ 0.70/L for the production of soy-based biodiesel was estimated (Haas et al. 2006). Palm oil is the main option that is traded internationally, and with potential for import in the short term (Dene and Hole 2006). Costs for production from palm oil have been estimated; the results are shown in Table 5.4 (Dene and Hole 2006).

5.2 Biofuels

Table 5.4 Costs of biodiesel production

Plant size (million liters)	Capital costs	Feedstock	Methanol	Other	Glycerol credit	Distribution and blending	Total
Tallow fat ($/L)							
6	0.33	0.40	0.05	0.11	0.12	0.08	0.85
23	0.15	0.40	0.05	0.10	0.12	0.08	0.66
46	0.11	0.40	0.05	0.09	0.12	0.09	0.61
69	0.08	0.40	0.05	0.08	0.12	0.15	0.64
137	0.06	0.40	0.05	0.06	0.12	0.15	0.60
Palm oil ($/L)							
60	0.09	0.73	0.05	0.08	0.12	0.04	0.88
71	0.08	0.73	0.05	0.08	0.12	0.04	0.86
143	0.06	0.73	0.05	0.06	0.12	0.04	0.82

The oil in vegetable seeds is converted into biodiesel through oil extraction, oil refining, and transesterification. The cost of biodiesel can be lowered by increasing feedstock yields, developing novel technologies, and increasing the economic return on glycerol production by finding other uses for this byproduct, which, at the moment, due to oversupply is sold for little or no value. Alternatively, the use of cosolvents, such as tetrahydrofuran, can consolidate the alcohol-oil-ester-glycerol system into a single phase, thereby reducing the processing costs (Granda *et al.* 2007). However, these improvements still would not make biodiesel economically competitive at the current stage.

Biofuel production costs can vary widely by feedstock, conversion process, scale of production, and region. On an energy basis, ethanol is currently more expensive to produce than gasoline in all regions considered. Only ethanol produced in Brazil comes close to competing with gasoline. Ethanol produced from corn in the US is considerably more expensive than from sugar cane in Brazil, and ethanol from grain and sugar beet in Europe is even more expensive. These differences reflect many factors, such as scale, process efficiency, feedstock costs, capital and labor costs, coproduct accounting, and the nature of the estimates.

The cost of large-scale production of bio-based products is currently high in developed countries. For example, the production cost of biofuels may be three times higher than that of petroleum fuels, without, however, considering the nonmarket benefits. Conversely, in developing countries, the costs of producing biofuels are much lower than in the OECD countries and very near to the world market price of petroleum fuel (UN 2006). Average international prices for common biocrude, fat, crops, and oils used as feedstock for biofuel production in 2007 are given in Table 5.5 (Demirbas 2008). The cost of feedstock is a major economic factor in the viability of biodiesel production. Nevertheless, the price of waste cooking oil is 2.5 to 3.5 times cheaper than virgin vegetable oils, and this can significantly reduce the total manufacturing cost of biodiesel (Table 5.5). Biodiesel obtained from waste cooking vegetable oils is considered a promising option. Waste cooking oil

Table 5.5 Average international prices for common biocrude, fat, crops, and oil used as feedstock for biofuel production in 2007 (US$/ton)

Biocrude	167
Maize	179
Sugar	223
Wheat	215
Crude palm oil	543
Rapeseed oil	824
Soybeen oil	771
Refined cottonseed oil	782
Crude corn oil	802
Crude peanut oil	891
Crude tea seed oil	514
Waste cooking oil	224
Yellow grease	412
Poultry fat	256

is available at relatively cheap prices for biodiesel production in comparison with fresh vegetable oil costs.

The economic advantages of a biofuel industry would include value added to feedstock, an increased number of rural manufacturing jobs, greater revenue from income taxes, investment in plant and equipment, reduced greenhouse gas emissions, reduction of a country's reliance on crude oil imports, and support for agriculture by providing new labor and market opportunities for domestic crops. In recent years, the importance of nonfood crops has increased significantly. The opportunity to grow nonfood crops under the compulsory set-aside scheme is an option to increase biofuel production.

Renewable liquid fuels such as bioethanol, biodiesel, green diesel, and green gasoline are important because they replace petroleum fuels. It is generally considered that renewable liquid fuels address many pressing concerns, including sustainability, reduction of greenhouse gas emissions, regional development, social structure and agriculture, and security of supply.

The socioeconomic impacts on the local economy arising from providing power through renewable resources instead of conventionally generated technologies are very important. These impacts include direct and indirect differences in jobs, income, and gross output. There are significant socioeconomic impacts associated with the investment in a new power plant, including increases in employment, output, and income in the local and regional economy. Increases in these categories occur as labor is directly employed in the construction and operation of a power plant, as local goods and services are purchased and utilized.

The potential for reduced costs of renewable liquid fuels and conservation of scarce fuel resources results in significant reductions in fuel usage. In addition to these economic benefits, development of renewable resources will have environmental, health, safety, and other benefits.

Agricultural ethanol is at present more expensive than synthesis-ethanol from ethylene. The simultaneous production of biomethanol (from sugar juice) in parallel to the production of bioethanol, appears economically attractive in locations where hydroelectricity is available at very low cost (~US$ 0.01/kWh) (RFA 2007).

Currently there is no global market for ethanol. The crop types, agricultural practices, land and labor costs, plant sizes, processing technologies, and government policies in different regions cause ethanol production costs and prices to vary considerably by region. The cost of producing bioethanol in a dry mill plant currently totals US$ 6.24/L. Corn accounts for 66% of operating costs while energy (electricity and natural gas) to fuel boilers and dry DDG represents nearly 20% of operating costs (Grassi 1999).

Ethanol from sugar cane, produced mainly in developing countries with warm climates, is generally much cheaper to produce than ethanol from grain or sugar beet in IEA countries. For this reason, in countries like Brazil and India, where sugar cane is produced in substantial volumes, sugar-cane-based ethanol is becoming an increasingly cost-effective alternative to petroleum fuels. Ethanol derived from cellulosic feedstock using enzymatic hydrolysis requires much greater processing than that derived from starch or sugar-based feedstock, but feedstock costs for grasses and trees are generally lower than for grain and sugar crops. If targeted reductions in conversion costs can be achieved, the total cost of producing cellulosic ethanol in OECD countries could fall below that of grain ethanol.

Estimates show that bioethanol in the EU becomes competitive when the oil price reaches US$ 70 a barrel while in the USA it becomes competitive at US$ 50 to $ 60 a barrel. For Brazil the threshold is much lower – between US$ 25 and US$ 30 a barrel. Other efficient sugar-producing countries such as Pakistan, Swaziland, and Zimbabwe have production costs similar to Brazil's (Urbanchuk 2007). Anhydrous ethanol, blendable with gasoline, is still somewhat more expensive. Prices in India have declined and are approaching the price of gasoline.

The generally larger US conversion plants produce biofuels, particularly ethanol, at lower cost than plants in Europe. Production costs for ethanol are much lower in countries with a warm climate, with Brazil probably the lowest-cost producer in the world. Production costs in Brazil, using sugar cane as the feedstock, have recently been recorded at less than half the costs in Europe. Production of sugar cane ethanol in developing countries could provide a low-cost source for substantial displacement of oil worldwide over the next 20 years.

For biofuels, the cost of crop feedstock is a major component of overall costs. In particular, the cost of producing oil-seed-derived biodiesel is dominated by the cost of the oil and by competition from high-value uses like cooking. The largest ethanol cost component is the plant feedstock. Operating costs, such as feedstock cost, co-product credit, chemicals, labor, maintenance, insurance, and taxes, represent about one third of total cost per liter, of which the energy needed to run the conversion facility is an important (and in some cases quite variable) component. Capital cost recovery represents about one sixth of the total cost per liter. It has been shown that plant size has a major effect on cost (Dufey 2006). The plant size can reduce operating costs by 15 to 20%, saving another $ 0.02 to $ 0.03/L. Thus, a large plant with production costs of $0.29/L may save $0.05 to $0.06/L over a smaller plant (Whims 2002).

Biodiesel from animal fat is currently the cheapest option ($0.4 to $0.5/L), while traditional transesterification of vegetable oil is at present around $0.6 to $0.8/L (IEA 2007). Rough projections of the cost of biodiesel from vegetable oil and waste grease are, respectively, $0.54 to $0.62/L and $0.34 to $0.42/L. With pretax diesel priced at $0.18/L in the USA and $0.20 to 0.24/L in some European countries, biodiesel is thus currently not economically feasible, and more research and technological development will be needed (Bender 1999).

5.2.4 Environmental Impacts of Biofuels

A number of studies with comparisons of diesel, natural gas, and diesel/biodiesel blend bus emissions have been published (Janulis 2004; Tzirakis *et al.* 2007; Krahl *et al.* 2009; Soltic *et al.* 2009; Coronado *et al.* 2009). Biodiesel has a good energy return because of the simplicity of its manufacturing process, has significant benefits in emissions as well, and could also play an important role in the energy economy if higher crop productivities are attained (Granda *et al.* 2007).

Table 5.6 shows the emissions of biodiesel (B20 and B100) for same model compression-ignition (diesel) vehicles (Demirbas 2009a). Emissions of NO_x increase with increasing biodiesel amounts in blends. The properties of biodiesel and diesel fuels, in general, show many similarities, and therefore biodiesel is rated as a realistic fuel as an alternative to diesel. There are several ways to control NO_x in a biodiesel engine. One could run the engine very lean, which lowers the temperature, or very rich, which reduces the oxygen supplies, decrease the burn time. Emissions of NO_x increase with the combustion temperature, the length of the high-temperature combustion period, and the availability of biodiesel, up to a point.

Alcohols have been used as a fuel for engines since the 19th century. Among the various alcohols, ethanol is known as the most suited renewable, bio-based, and eco-friendly fuel for spark-ignition (SI) engines. The most attractive properties of ethanol as an SI engine fuel are that it can be produced from renewable energy

Table 5.6 Emissions of biodiesel for same model diesel vehicles

Vehicle or Engine	Fuel	Emissions, g/km				
		NO_x	CO	CH	PM	SO_x
Peugeot Partner	B100	2.05	9.37	0.54	2.68	0
Peugeot Partner	B20	1.86	17.73	1.32	4.71	0.004
Renault Kangoo	B100	2.23	9.22	0.49	3.06	0
Renault Kangoo	B20	1.92	17.36	1.26	5.63	0.003
Dacia Pickup	B100	2.15	9.42	0.56	2.59	0
Dacia Pickup	B20	1.91	18.29	1.35	4.63	0.005

5.2 Biofuels

sources such as sugar, cane, cassava, many types of waste biomass materials, corn, and barley. In addition, ethanol has a higher evaporation heat, octane number, and flammability temperature; therefore it has a positive influence on engine performance and reduces exhaust emissions. The results of an engine test showed that ethanol addition to unleaded gasoline increases engine torque, power, and fuel consumption and reduces carbon monoxide (CO) and hydrocarbon emissions (Demirbas 2009a).

The biodiesel impacts on exhaust emissions vary depending on the type of biodiesel and on the type of conventional diesel. Blends of up to 20% biodiesel mixed with petroleum diesel fuels can be used in nearly all diesel equipment and are compatible with most storage and distribution equipment. Using biodiesel in a conventional diesel engine substantially reduces emissions of unburned hydrocarbons, carbon monoxide, sulfates, polycyclic aromatic hydrocarbons, nitrated polycyclic aromatic hydrocarbons, and particulate matter. These reductions increase as the amount of biodiesel blended into diesel fuel increases. In general, biodiesel increases NO_x emissions when used as fuel in a diesel engine. The fact that NO_x emissions increase with increasing biodiesel concentration could be a detriment in areas that where ozone forms. The pollutant emissions of ethanol–gasoline blends of 0, 5, 10, 15, and 20% were experimentally analyzed in a four-stroke (SI) engine. The concentration of CO and HC emissions in the exhaust pipe were measured and found to decrease when ethanol blends were introduced. This was due to the high oxygen percentage in the ethanol. In contrast, the concentration of CO_2 and NO_x was found to increase when ethanol was introduced (Najafi *et al.* 2009).

Oxygenated diesel fuel blends have the potential to reduce the emission of particulate matter and to be an alternative to diesel fuel. Results obtained showed that the addition of bioethanol to the diesel fuel may be necessary to decrease diesel particulate matter generation during combustion (Corro and Ayala 2008; Yu and Tao 2009). The total number and total mass of the particulate matter of ethanol–diesel blend fuels were decreased by about 11.7 to 26.9% (Kim and Choi 2008).

An experimental investigation was conducted to evaluate the effects of using blends of ethanol with conventional diesel fuel, with 5 and 10% (by vol.) ethanol, on the performance and exhaust emissions of a fully instrumented, six-cylinder, turbocharged and after-cooled, heavy-duty, direct-injection, Mercedes–Benz engine. Fuel consumption, exhaust smokiness, and exhaust-regulated gas emissions such as nitrogen oxides, carbon monoxide, and total unburned hydrocarbons were measured. The differences in the measured performance and exhaust emissions of the two ethanol–diesel fuel blends from the baseline operation of the engine, i.e., when working with neat diesel fuel, were determined and compared (Rakopoulos *et al.* 2008). Diesel emissions were measured from an automotive engine using anhydrous bioethanol blended with conventional diesel, with 10% ethanol in volume and no additives. The resulting emissions were compared with those from pure diesel (Lapuerta *et al.* 2008)

The results of the statistical analysis suggest that the use of E10 results in statistically significant decreases in CO emissions (-16%); statistically significant increases in emissions of acetaldehyde (108%), 1,3-butadiene (16%), and benzene (15%); and no statistically significant changes in NO_x, CO_2, CH_4, N_2O, or formaldehyde emissions. The statistical analysis suggests that the use of E85 results in statistically significant decreases in emissions of NO_x (-45%), 1,3-butadiene (-77%), and benzene (-76%); statistically significant increases in emissions of formaldehyde (73%) and acetaldehyde (2,540%), and no statistically significant change in CO and CO_2 emissions (Graham *et al.* 2008).

Biofuels are important because they replace petroleum fuels. There are many benefits for the environment, economy, and consumers in using biofuels. The advantages of biofuels such as biodiesel, vegetable oil, bioethanol, biomethanol, biomass pyrolysis oil as engine fuel are liquid nature-portability, ready availability, renewability, higher combustion efficiency, lower sulfur and aromatic content and biodegradability. The biggest difference between biofuels and petroleum feedstocks is oxygen content. Biofuels have oxygen levels of 10 to 45%, while petroleum has essentially none, making the chemical properties of biofuels very different from those of petroleum. Oxygenates are just preused hydrocarbons having a structure that provides a reasonable antiknock value. Also, as they contain oxygen, fuel combustion is more efficient, reducing hydrocarbons in exhaust gases. The only disadvantage is that oxygenated fuel has less energy content.

Combustion is the chemical reaction of a particular substance with oxygen. It is a chemical reaction during which from certain matters other simple matters are produced. This is a combination of inflammable matter with oxygen from the air accompanied by heat release. The quantity of heat involved when one mole of a hydrocarbon is burned to produce carbon dioxide and water is called the heat of combustion. Combustion to produce carbon dioxide and water is characteristic of organic compounds; under special conditions it is used to determine their carbon and hydrogen content. During combustion the combustible part of fuel is subdivided into volatile parts and solid residue. During heating it evaporates together with some carbon in the form of hydrocarbons, combustible gases, and carbon monoxide release by thermal degradation of the fuel. Carbon monoxide is mainly formed by the following reactions: first from a reduction in CO_2 with unreacted C,

$$CO_2 + C \rightarrow 2CO \qquad (5.2)$$

and, second, from the degradation of carbonyl fragments (–CO) in fuel molecules at temperatures of 600 to 750 K.

The combustion process is started by heating the fuel above its ignition temperature in the presence of oxygen or air. Under the influence of heat, the chemical bonds of the fuel are cleaved. If complete combustion occurs, the combustible elements (C, H, and S) react with the oxygen content of the air to form CO_2, H_2O, and, mainly, SO_2.

If insufficient oxygen is present or the fuel and air mixture is insufficient, then the burning gases are partially cooled below the ignition temperature and the combustion process stays incomplete. The flue gases then still contain combustible components, mainly carbon monoxide (CO), unburned carbon (C), and various hydrocarbons ($C_x H_y$).

5.2 Biofuels

The standard measure of the energy content of a fuel is its heating value (HV), sometimes called the calorific value or heat of combustion. In fact, there are multiple values for the HV, depending on whether it measures the enthalpy of combustion (ΔH) or the internal energy of combustion (ΔU), and whether for a fuel containing hydrogen product water is accounted for in the vapor phase or the condensed (liquid) phase. With water in the vapor phase, the lower heating value (LHV) at constant pressure measures the enthalpy change due to combustion (Jenkins *et al.* 1998). The HV is obtained by the complete combustion of a unit quantity of solid fuel in an oxygen-bomb colorimeter under carefully defined conditions. The gross heat of combustion or higher heating value (GHC or HHV) is obtained by the oxygen-bomb colorimeter method as the latent heat of moisture in the combustion products is recovered.

5.2.5 Combustion Efficiencies of Biofuels

Biofuels are oxygenated compounds. Oxygenated compounds such as ethanol, methanol, and biodiesel provide more efficient combustion and cleaner emissions. At a stoichiometric air:fuel ratio of 9:1 in comparison with gasoline's 14.7:1, it is obvious that more ethanol is required to produce the chemically correct products of CO_2 and water.

Ethanol has a higher octane number (108), broader flammability limits, higher flame speeds, and higher heats of vaporization than gasoline. These properties allow for a higher compression ratio, shorter burn time, and leaner burn engine, which lead to theoretical efficiency advantages over gasoline in an internal combustion engine. The octane number of ethanol allows it to sustain significantly higher internal pressures than gasoline, before being subjected to predetonation.

Disadvantages of ethanol include its lower energy density than gasoline, its corrosiveness, low flame luminosity, lower vapor pressure, miscibility with water, and toxicity to ecosystems.

Methanol also allows one to take advantage of the higher octane number of methyl (114) alcohol and increase the engine compression ratio. This would increase the efficiency of converting the potential combustion energy to power. Finally, alcohols burn more completely, thus increasing combustion efficiency. Some technical properties of fuels are presented in Table 5.7.

Table 5.7 Some technical properties of fuels

Fuel property	Gasoline	No. 2 diesel	Isoctane	Methanol	Ethanol
Cetane number	–	50	–	5	8
Octane number	96	–	100	114	108
Autoignition temperature (K)	644	588	530	737	606
Lower heating value (MJ/Kg)	44	43	45	20	27

Biofuels such as bioethanol, biomethanol, biohydrogen, and biodiesel generally have lower emissions than fossil-based engine fuels. Many studies on the performances and emissions of compression ignition engines, fueled with pure biodiesel and blends with diesel oil, have been performed and are reported in the literature (Laforgia and Ardito 1994; Cardone *et al.* 1998).

Vegetable oils have become more attractive recently because of their environmental benefits and the fact that they is made from renewable resources. Dorado *et al.* (2003) describe experiments on the exhaust emissions of biodiesel from olive oil methyl ester as alternative diesel fuel in a diesel direct injection Perkins engine.

The methyl ester of vegetable oil was evaluated as a fuel in CIE by researchers (Dunn 2001), who concluded that the performance of the esters of vegetable oil did not differ greatly from that of diesel fuel. The brake power was nearly the same as with diesel fuel, while the specific fuel consumption was higher than that of diesel fuel. Based on crankcase oil analysis, engine wear rates were low but some oil dilution did occur. Carbon deposits inside the engine were normal, with the exception of intake valve deposits.

The results showed that transesterification treatment decreased the injector coking to a level significantly lower than that observed with diesel fuel (Shay 1993). Although most researchers agree that vegetable oil ester fuels are suitable for use in CIE, a few contrary results have also been obtained. The results of these studies point out that most vegetable oil esters are suitable as diesel substitutes but that more long-term studies are necessary for commercial utilization to become practical.

The use of biodiesel to reduce N_2O is attractive for several reasons. First, biodiesel contains little nitrogen, as compared with diesel fuel, which is also used as a reburning fuel. The N_2O reduction is strongly dependent on initial N_2O concentration and only slightly dependent upon temperature, where increased temperature increased N_2O reduction. This results in lower N_2O production from fuel nitrogen species for biodiesel. In addition, biodiesel contains a virtually trace amount of sulfur, so SO_2 emissions are reduced in direct proportion to the diesel fuel replacement. Bioesel is a regenerable fuel; when a diesel fuel is replaced by a biodiesel, there is a net reduction in CO_2 emissions. As an energy source used in a diesel engine it reduces the consumption of diesel fuels and thereby reduces the greenhouse effect. Additional effects are a reduction in the ash volume and the SO_x and NO_x emissions of a biodiesel-fueled CIEs. Neat BD and BD blends reduce particulate matter, hydrocarbons (HC), and carbon monoxide (CO) emissions and increase nitrogen oxide (NO_x) emissions compared with diesel fuel used in an unmodified diesel engine (Grassi 1999).

The total net emission of carbon dioxide (CO_2) from biodiesel is considerably less than that of diesel oil, and the amount of energy required for the production of biodiesel is less than that obtained with the final product of diesel oil. In addition, the emission of pollutants is somewhat less. CO_2, one of the primary greenhouse gases, is a transboundary gas, which means that, after being emitted by a source, it is quickly dispersed in the atmosphere by natural processes. Table 5.8 shows the average biodiesel emissions compared to conventional diesel, according to EPA (2002).

5.2 Biofuels

Table 5.8 Average biodiesel emissions (%) compared to conventional diesel

Emission type	Pure biodiesel	20% biodiesel + 80% petrodiesel
	B100	B20
Total unburned hydrocarbons (HC)	−67	−20
Carbon monoxide	−48	−12
Particulate matter	−47	−12
NO_x	+10	+2
Sulfates	−100	−20
Polycyclic aromatic hydrocarbons[b]	−80	−13
Ozone potential of speciated HC	−50	−10

[a] Estimated from B100 result.
[b] Average reduction across all compounds measured.

Table 5.9 Average changes in mass emissions from diesel engines using biodiesel mixtures relative to standard diesel fuel (%)

Mixture	CO	NO_x	SO_2	Particular matter	Volatile organic compounds
B20	−13.1	+2.4	−20	−8.9	−17.9
B100	−42.7	+13.2	−100	−55.3	−63.2

Table 5.9 shows the average changes in mass emissions from diesel engines using biodiesel mixtures relative to standard diesel fuel.

Results indicate that the transformities of biofuels are greater than those of fossil fuels, thus showing that more resources are required to obtain the environmental friendly product. This can be explained by the fact that natural processes are more efficient than industrial ones. On the other hand, the time involved in the formation of fossil fuels is considerably different from that required for the production of biomass (Carraretto *et al.* 2004).

Different scenarios for the use of agricultural residues as fuel for heat or power generation have been analyzed. Reductions in net CO_2 emissions are estimated at 77 to 104 g/MJ of diesel displaced by biodiesel. The predicted reductions in CO_2 emissions are much greater than values reported in recent studies on biodiesel derived from other vegetable oils, due both to the large amount of potential fuel in the residual biomass and to the low-energy inputs in traditional coconut farming techniques. Unburned hydrocarbon emissions from biodiesel fuel combustion decrease compared to regular petroleum diesel.

Biodiesel, produced from different vegetable oils, seems very interesting for several reasons: it can replace diesel oil in boilers and internal combustion engines without major adjustments; only a small decrease in performance has been reported; it produces almost zero emissions of sulfates; it generates a small net contribution of CO_2 when the whole life cycle is considered; the emission of pollutants is comparable with that of diesel oil. For these reasons, several campaigns have been planned in many countries to introduce and promote the use of biodiesel (Carraretto *et al.* 2004).

The brake power of biodiesel was nearly the same as with diesel fuel, while the specific fuel consumption was higher than that of diesel fuel. Based on crankcase oil analysis, engine wear rates were low but some oil dilution did occur. Carbon deposits inside the engine were normal, with the exception of intake valve deposits.

Several results related to the influence of micronutrition on oil yields have been reported (Nithedpattrapong *et al.* 1995; Tilman *et al.* 2006; Thamsiriroj 2007; Thamsiriroj and Murphy 2009). Suitable climatic and soil conditions have increased plant oil yields (Thamsiriroj 2007). Biomethane generated from grass requires four times less land than biodiesel from rape seed to produce the same gross energy (Thamsiriroj and Murphy 2009). Grass is a low-energy, carbon-negative input crop (Tilman *et al.* 2006; Thamsiriroj and Murphy 2009).

Biofuel consumption in the EU is growing rapidly, but major efforts will need to be undertaken if the EU's objectives for 2010 and beyond are to be achieved. Current and future policy support therefore focuses on creating favorable economic or legal frameworks to accelerate the market penetration of biofuels. The EU member states will promote specific types of biofuels, depending on their main objectives and natural potentials. The main EU directives that have an impact on sustainable energy development are those promoting energy efficiency and use of renewable energy sources, those implementing greenhouse gas mitigation and atmospheric pollution reduction policies, and other policy documents and strategies targeting the energy sector. Promotion of use of renewable energy sources and energy efficiency improvements are among the priorities of the EU's energy policy because the use of renewable energy sources and energy efficiency improvements has a positive impact on energy security and climate change mitigation. In the EU, climate change has been the principal policy driver for promoting the use of energy from renewable resources. The spring European Council of 2007 set ambitious targets by 2020 of a 20% reduction in greenhouse gas emissions compared to 1990 levels and a 20% renewables share in the EU's final energy consumption, including a 10% share of biofuels in each member state. The greenhouse gas emission savings from the use of biofuels will be at least 35% compared to fossil fuels (Demirbas 2009b).

On 23 January 2008, the Proposal for a Directive of the European Parliament and of the Council on the Promotion of the Use of Energy from Renewable Resources was issued. Article 15 of this proposal defined environmental sustainability criteria that both domestic and imported biofuels must satisfy. These criteria include a minimum greenhouse-gas-reduction requirement, limits on the types of land where conversion into biofuel crop production is acceptable, and reinforcement of best agricultural practices. The first sustainability criterion defined in the European proposal is a 35% reduction in greenhouse gas emissions for biofuels relative to their fossil fuel counterpart. The second sustainability criterion relates to the preservation of diverse ecosystems. The greenhouse gas emission savings from the use of biofuels and other bioliquids shall be at least 35%. Biofuels and other bioliquids shall not be made from raw material obtained from highly biodiverse land and from land with high carbon stock. Agricultural raw materials cultivated in the European Community and used for the production of biofuels and other bioliquids shall be obtained in accordance with the requirements and standards under the provisions of the proposal, and in accordance with the minimum requirements for good agricultural and environmental conditions (Demirbas 2009b).

Specifically for biofuels and bioliquids the draft directive establishes environmental sustainability criteria to ensure that biofuels that are to count towards the target are sustainable. They must achieve a minimum level of greenhouse gas emission reduction (35%) and respect a number of binding requirements related to environmental impact and biodiversity. The sustainability criteria aim to reduce greenhouse gas emissions and to prevent loss of valuable biodiversity and undesired land use changes. For the foreseeable future the EU will have to rely on first-generation biofuels to achieve the 10% target by 2020: vegetable oils for biodiesel and sugar and starch crops for bioethanol. These import requirements may be between 30 and 50%. The sustainability criteria under discussion in the EU constitute an important step forward, their shortcomings notwithstanding. They should be implemented in a nondiscriminatory way without conferring undue competitive advantage to domestic producers and allowing developing countries export opportunities for bioethanol and biomass. A speedy transition to second-generation biofuels is of particular importance. Sustainability criteria should progressively ensure that only advanced biofuels are available for end users (Demirbas 2009b).

5.2.6 Bio-oil

The term bio-oil is used mainly to refer to liquid fuels. When biomass is processed under high temperature in the absence of oxygen, products are produced in three phases: the vapor phase, the liquid phase, and the solid phase. The liquid phase is a complex mixture called bio-oil. The compositions of bio-oils vary significantly with the types of feedstock and processing conditions.

Thermochemical conversion is a process through which biomass in the absence of oxygen and at high temperature can be converted into various fuels including char, oil, and gas. The resulting bio-oils present an alternative to liquid biofuels with similarities to petroleum oil (Kishimoto *et al.* 1994). The process can be subdivided into pyrolysis and thermochemical liquefaction (Demirbas 2000).

Bio-oils are liquid or gaseous fuels made from biomass materials, such as agricultural crops, municipal wastes, and agricultural and forestry byproducts, *via* biochemical or thermochemical processes.

Bio-oil has a higher energy density than biomass and can be obtained by quick heating of dried biomass in a fluidized bed followed by cooling. The byproduct char and gases can be combusted to heat the reactor. For utilization of biomass in remote locations, it is more economical to convert the biomass into bio-oil and then transport the bio-oil. Bio-oil can be used in vehicle engines – either totally or partially in a blend.

Biomass is dried and then converted into an oily product known as bio-oil by very quick exposure to heated particles in a fluidized bed. The char and gases produced are combusted to supply heat to the reactor, while the product oils are cooled and condensed. The bio-oil is then shipped by truck from these locations to the hydrogen production facility. It is more economical to produce bio-oil at remote locations and then ship the oil, since the energy density of bio-oil is higher than biomass. For this analysis, it was assumed that the bio-oil would be produced at several smaller plants that are closer to the sources of biomass so that lower-cost feedstocks could be obtained.

The feasibility of producing liquid fuel or bio-oil via pyrolysis or thermochemical liquefaction of microalgae has been demonstrated for a range of microalgae (Dote et al. 1994; Sawayama et al. 1999; Peng et al. 2000, 2001; Tsukahara and Sawayama 2005; Demirbas 2006).

Five moss samples (*Polytrichum commune, Dicranum scoparium, Thuidium tamarascinum, Sphagnum palustre, Drepanocladus revolvens*), one alga sample (*Cladophora fracta*), and one microalga sample (*Chlorella protothecoides*) were used in the earlier work (Demirbas 2006). The yields of bio-oil from the samples via pyrolysis are presented as a function of temperature (K) in Figure 5.3. The yield of bio-oil from pyrolysis of the samples increased with temperature, as expected. The yields were increased up to 750 K in order to reach the plateau values at 775 K. The maximum yields were 39.1, 34.3, 33.6, 37.0, 35.4, 48.2, and 55.3% of the sample for *Polytrichum commune, Dicranum scoparium, Thuidium tamarascinum, Sphag-*

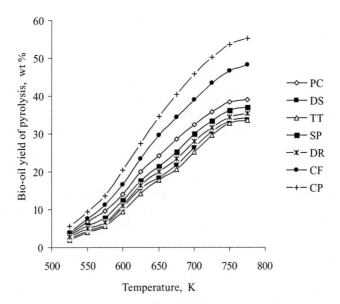

Figure 5.3 Yields of bio-oil from pyrolysis of the samples: *Polytrichum commune* (PC), *Dicranum scoparium* (DS), *Thuidium tamarascinum* (TT), *Sphagnum palustre* (SP), *Drepanocladus revolvens* (DR), *Cladophora fracta* (CF), and *Chlorella protothecoides* (CP)

5.2 Biofuels

num palustre, Drepanocladus revolvens, Cladophora fracta and Chlorella protothe-coides, respectively.

5.2.7 Biomethane

The concept of using algae as a fuel was first proposed by Meier (1955) for the production of methane gas from the carbohydrate fraction of cells. This idea was further developed by Oswald and Golueke (1960), who introduced a conceptual technoeconomic engineering analysis of digesting microalgal biomass grown in large raceway ponds to produce methane gas. In the 1970s, as the cost of conventional fuels began rising rapidly, the possibility of using algae as a fuel source received renewed attention. A more detailed design and engineering analysis of this concept was carried out by Benemann *et al.* (1978), who concluded that such systems could produce biogas competitively with projected fossil fuel prices.

Anaerobic digestion of biowastes occurs in the absence of air, and the resulting gas, known as biogas, is a mixture consisting mainly of methane and carbon dioxide. Biogas is a valuable fuel that is produced in digesters filled with feedstock like dung or sewage. The digestion is allowed to continue for a period of 10 d to a few weeks.

Algal biomass can be used for biogas production. In Poland, there are numerous active biogas installations, from large-scale ones to small ones fed with straw and green plant fuel that serve a few farms; so far, however, algae have not been used as a fuel. Some macroalgal species like *Macrocystis pylifera* and genera such as *Sargassum, Laminaria, Ascophyllum, Ulva, Cladophora, Chaetomorpha,* and *Gracilaria* have been explored as potential methane sources (Filipkowska *et al.* 2008). But in spite of the large seaweed biomass in various regions of the world, anaerobic digestion for biogas generation appears to be unsatisfactory and therefore uneconomical (Gunaseelan 1997; Caliceti *et al.* 2002).

Anaerobic digestion of algal waste produces carbon dioxide, methane, and ammonia. Leftover nitrogen and phosphorus compounds can be reused as fertilizer for the algal process. Using methane as an energy source can further enhance energy recovery from the process.

Researchers have highlighted some key issues to be addressed in microalgal production:

- Sodium (in salt) can inhibit the anaerobic digestion process when using marine algae, although researchers say that suitable bacteria (anaerobic digesters) can adapt.
- Digestion of algae can be enhanced and the methane yield increased by physical or chemical pretreatment to break down cell walls and make the organic matter in the cells more accessible.
- The nitrogen content of certain algae can be high, resulting in greater levels of ammonia, which can also inhibit the digestion process. One strategy to overcome this problem uses a "codigestion" process, whereby other organic waste, which is higher in carbon and lower in nitrogen, is added to the algal waste.

5.2.8 Production of Bio-oil and Hydrogen by Pyrolysis

Hydrogen is an important fuel with wide applications in fuel cells, liquefaction of coal, and upgrading of heavy oils (e.g., bitumen). Hydrogen can be produced biologically by a variety of means, including the steam reformation of bio-oils, dark and photo fermentation of organic materials, and photolysis of water catalyzed by special microalgal species.

The chemical compositions of algae are given in Table 5.10 (Demirbas 2007). Algae are mainly composed of proteins, lipids, and water-soluble carbohydrates.

Two moss samples (*Polytrichum commune*, *Thuidium tamarascinum*), one alga sample (*Cladophora fracta*), and one microalga sample (*Chlorella protothecoides*) were subjected to pyrolysis and steam gasification for producing hydrogen-rich gas (Demirbas, unpublished work).

The temperature of the reaction vessel was measured with an iron-constantan thermocouple and controlled at ± 3 K. The pyrolysis experiments were performed at temperatures of 575, 625, 675, 725, 775, 825, and 925 K. The steam gasification experiments were carried out at temperatures of 825, 875, 925, 975, 1,025, 1,075, 1,125, 1,175, and 1,225 K (Demirbas, unpublished work).

Table 5.11 shows the proximate analysis data and higher heating values (HHVs) of samples. The HHV (MJ/kg) of the moss and alga samples as a function of fixed carbon (FC) wt% can be calculated from

$$HHV = 0.322 \, (FC) + 10.7123 \tag{5.3}$$

The HHVs can be calculated using Equation 5.3 and represent high correlation obtained by means of regression analysis. The correlation coefficient r is 0.999.

Table 5.10 Chemical compositions of algae on a dry matter basis (%)

Species of sample	Proteins	Carbohydrates	Lipids	Nucleic acid
Scenedesmus obliquus	50–56	10–17	12–14	3–6
Scenedesmus quadricauda	47	–	1.9	–
Scenedesmus dimorphus	8–18	21–52	16–40	–
Chlamydomonas rheinhardii	48	17	21	–
Chlorella vulgaris	51–58	12–17	14–22	4–5
Chlorella pyrenoidosa	57	26	2	–
Spirogyra sp.	6–20	33–64	11–21	–
Dunaliella bioculata	49	4	8	–
Dunaliella salina	57	32	6	–
Euglena gracilis	39–61	14–18	14–20	–
Prymnesium parvum	28–45	25–33	22–38	1–2
Tetraselmis maculata	52	15	3	–
Porphyridium cruentum	28–39	40–57	9–14	–
Spirulina platensis	46–63	8–14	4–9	2–5
Spirulina maxima	60–71	13–16	6–7	3–4.5
Synechoccus sp.	63	15	11	5
Anabaena cylindrica	43–56	25–30	4–7	–

5.2 Biofuels

Table 5.11 Proximate analysis data and higher heating values (HHVs) of samples (% dry wt basis)

Sample	Fixed carbon	Volatile matter	Ash	HHV (MJ/kg)
Polytrichum commune	19.4	65.8	14.8	17.0
Thuidium tamarascinum	15.4	72.3	12.3	15.5
Cladophora fracta	28.1	65.6	6.3	19.8
Chlorella protothecoides	39.6	54.6	5.8	23.6

The yields of bio-oil from the samples via pyrolysis are presented as a function of temperature (K) in Figure 5.4. The yield of bio-oil from pyrolysis of the samples increased with temperature, as expected. The yields were increased up to 750 K in order to reach the plateau values at 775 K. The maximum yields for *Polytrichum commune*, *Thuidium tamarascinum*, *Cladophora fracta*, and *Chlorella protothecoides* were 31.6, 37.3, 45.0, and 50.8% of the sample at 925 K, respectively. The bio-oil yields of pyrolysis from algae were higher than those of mosses. Bio-oil comparable to fossil oil was obtained from microalgae (Miao and Wu 2004). In the pyrolysis process, the yield of charcoal decreases with increasing pyrolysis temperature. The yield of the liquid product is highly excessive at temperatures between 625 and 725 K.

The HHVs for bio-oils from mosses 21.5 to 24.8 MJ/kg and the HHVs for bio-oils from algae and microalgae 32.5 and 39.7 MJ/kg, respectively, were obtained by pyrolysis at temperatures ranging from 775 to 825 K. In general, algae bio-oils are of a higher quality than bio-oils from mosses.

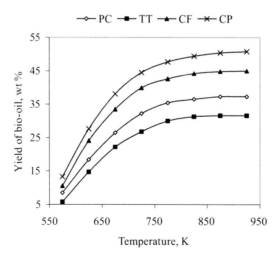

Figure 5.4 Yield of bio-oil from moss and alga samples by pyrolysis at different temperatures (K). *Polytrichum commune* (PC), *Thuidium tamarascinum* (TT), *Cladophora fracta* (CF), and *Chlorella protothecoides* (CP)

Figure 5.5 shows the effect of temperature on yields of gaseous products from the samples by pyrolysis. As can be seen in Figure 5.4, the yields of gaseous products from the samples of *Polytrichum commune*, *Thuidium tamarascinum*, *Cladophora fracta*, and *Chlorella protothecoides* increased from 5.3 to 40.6%, 6.5 to 42.2%, 8.2 to 39.2%, and 9.5 to 40.6% by volume, respectively, while the final pyrolysis temperature was increased from 575 to 875 K.

Figure 5.6 shows the plots for yields of hydrogen in gaseous products from the samples by pyrolysis. The percent of hydrogen in gaseous products from the samples of *Polytrichum commune*, *Thuidium tamarascinum*, *Cladophora fracta*, and *Chlorella protothecoides* increased from 21.3 to 38.7%, 23.0 to 41.3%, 25.8 to 44.4%, and 27.6 to 48.7% by volume, respectively, while the final pyrolysis temperature was increased from 650 to 875 K.

Figure 5.7 shows the plots for yields of hydrogen in gaseous products from the samples by steam gasification. The percent of hydrogen in gaseous products from the samples of *Polytrichum commune*, *Thuidium tamarascinum*, *Cladophora fracta*, and *Chlorella protothecoides* increased from 21.8 to 50.0%, 23.5 to 52.0%, 26.3 to 54.7%, and 28.1 to 57.6% by volume, respectively, while the final gasification temperature was increased from 825 to 1,225 K.

Figure 5.8 shows the plots for yields of hydrogen in gaseous products from microalgae and wood samples by pyrolysis. The percent of hydrogen in gaseous products from the samples of beech wood and spruce wood increased from 31.5 to 40.5% and 33.3 to 42.3% by volume, respectively, while the final pyrolysis temperature was increased from 650 to 875 K (Demirbas and Arin 2004). Microalgae gaseous products are higher quality than gaseous products from wood (Figure 5.8). In general, algal gaseous products are of higher quality than gaseous products from mosses.

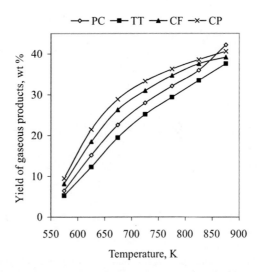

Figure 5.5 Yields of gaseous products from the samples by pyrolysis: *Polytrichum commune* (PC), *Thuidium tamarascinum* (TT), *Cladophora fracta* (CF), and *Chlorella protothecoides* (CP)

5.2 Biofuels

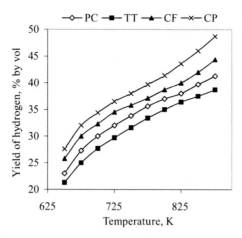

Figure 5.6 Yields of hydrogen in gaseous products from the samples by pyrolysis: *Polytrichum commune* (PC), *Thuidium tamarascinum* (TT), *Cladophora fracta* (CF), and *Chlorella protothecoides* (CP)

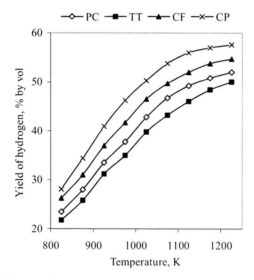

Figure 5.7 Yields of hydrogen in gaseous products from the samples by steam gasification: *Polytrichum commune* (PC), *Thuidium tamarascinum* (TT), *Cladophora fracta* (CF), and *Chlorella protothecoides* (CP)

Table 5.12 shows the yields of bio-oil by pyrolysis from moss and algae samples (Demirbas 2006). As can be seen from Table 5.12, the bio-oil yield for *Chlorella protothecoides* (a microalgae sample) rose from 12.8 to 55.3% as the temperature rose from 575 to 775 K, and then gradually decreased to 51.8% was obtained at 875 K with a heating rate of 10 K/s. The bio-oil yield for *Polytrichum commune* (a moss sample) rose from 10.3 to 39.1% as the temperature rose from 575 to 775 K,

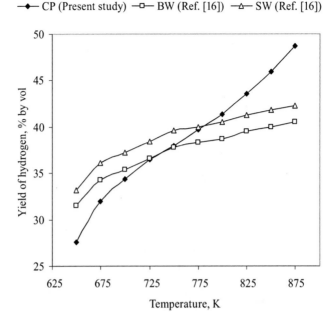

Figure 5.8 Yields of hydrogen in gaseous products from microalgae and wood samples by pyrolysis: *Chlorella protothecoides* (CP), beech wood (BW), and spruce wood (SW)

Table 5.12 Yields of bio-oil by pyrolysis from moss and algae samples at different temperatures (K)

Sample	575	625	675	725	775	825	875
Polytrichum commune	10.3	20.0	27.5	35.8	39.1	38.4	36.7
Dicranum scoparium	6.0	15.5	21.8	30.7	34.3	33.8	31.7
Thuidium tamarascinum	5.6	14.2	20.7	29.5	33.6	33.4	31.3
Sphagnum palustre	7.9	17.7	25.3	33.5	37.0	36.3	34.6
Drepanocladus revolvens	6.7	16.4	23.5	31.7	35.4	34.7	32.9
Cladophora fracta	10.5	23.5	33.2	43.4	48.2	46.8	44.6
Chlorella protothecoides	12.8	27.4	38.4	50.2	55.3	53.7	51.6

and then gradually decreased to 36.7% was obtained at 875 K with a heating rate of 10 K/s (Demirbas 2006). For algae, maximum bio-oil yields of between 48.2 and 46.8%, and for microalgae 55.3 and 53.7% were obtained at temperatures ranging from 775 to 825 K, whereas for wood, cotton stalk, tobacco stalk, and sunflower bagasse, maximum oil yields between 39.7 and 49.4% were obtained at temperatures ranging from 775 to 825 K (Putun 2002; Gercel 2002).

Table 5.13 shows the yields of gaseous product by pyrolysis from moss and algae samples (Demirbas 2006). As shown in Tables 5.12 and 5.13, the yields of gaseous products for *Chlorella protothecoides* increased from 9.5 to 39.5% as the tempera-

5.2 Biofuels

Table 5.13 Yields of gaseous product by pyrolysis from moss and algae samples at different temperatures (K)

Sample	575	625	675	725	775	825	875
Polytrichum commune	6.5	14.8	22.6	26.4	29.2	36.6	42.2
Dicranum scoparium	5.8	12.5	19.8	25.0	27.6	35.0	40.8
Thuidium tamarascinum	5.3	11.2	17.9	23.5	25.6	33.2	39.3
Sphagnum palustre	5.5	11.9	18.3	24.2	26.5	34.0	39.8
Drepanocladus revolvens	5.6	12.3	18.9	24.7	27.0	34.5	40.4
Cladophora fracta	8.2	19.7	28.2	32.6	35.7	38.0	39.7
Chlorella protothecoides	9.5	21.8	29.5	33.7	36.3	38.1	39.5

ture rose from 575 to 875 K. The char yields of pyrolysis from mosses were higher than those of algae.

With the interaction of water and char from decomposition of biomass occur intermediate products, which leads to more hydrogen-rich gas yield by steam reforming. The pyrolysis is carried out at moderate temperatures and steam gasification at the highest temperatures. In order to clarify the steam gasification mechanism in detail, more kinetic study is necessary. These results suggest that the fundamental information obtained in the gasification of each component could possibly be used to predict the composition of product gas generated in air-steam gasification of biomass.

5.2.9 Anaerobic Biohydrogen Production

In the microbial fermentation of biomass, different waste materials can be used as substrates. A new and unique process has been developed where substrates such as carbohydrates are fermented by a consortium of bacteria; they produce hydrogen and carbon dioxide. Highly concentrated organic waste water is one of the most abundantly available biomasses that can be exploited for microbial conversion into hydrogen (Nath and Das 2003). Municipal solid wastes and digested sewage sludge have the potential to produce large amounts of hydrogen by suppressing the production of methane through the introduction of low-voltage electricity into the sludge. The substrate from the acidogenesis of fruit and vegetable market wastes gives higher hydrogen evolution rates (about threefold) compared to synthetic medium. A mixed culture of photosynthetic anaerobic bacteria provides a method of utilization of a variety of resources for biohydrogen production (Miyake *et al.* 1990).

Hydrogen produced by photosynthetic organisms is one of a range of popular scenarios for renewable energy. Hydrogen can be produced by algae under specific conditions. Three different ways to produce hydrogen have been proposed: direct and indirect photolysis and ATP-driven hydrogen production. Direct photolysis is possible when the resulting hydrogen and oxygen are continuously flushed away.

126   5 Energy from Algae

Photosynthetic water splitting are coupled, results in the simultaneous production of hydrogen and oxygen. This results in major safety risk and costs to separate the hydrogen and oxygen. Major factors affecting the cost of hydrogen production by microalgae are the cost of the huge photobioreactor and the cost of hydrogen storage facilities that guarantee continuous hydrogen supply both during the night or during cloudy periods of the day.

Anaerobic hydrogen production proceeds photofermentatively as well as without the presence of light. Anaerobic bacteria use organic substances as the sole source of electrons and energy, converting them into hydrogen.

$$\text{Glucose} + 2H_2O \rightarrow 2\text{Acetate} + 2CO_2 + 4H_2 \tag{5.4}$$

$$\text{Glucose} \rightarrow \text{Butyrate} + 2CO_2 + 2H_2 \tag{5.5}$$

The reactions involved in hydrogen production (Equations 5.4 and 5.5) are rapid, and these processes do not require solar radiation, making them useful for treating large quantities of wastewater by using a large fermentor.

Since they cannot utilize light energy, the decomposition of organic substrates is incomplete. Further decomposition of acetic acid is not possible under anaerobic conditions. Nevertheless, these reactions are still suitable for the initial steps of wastewater treatment and hydrogen production followed by further waste treatment stages.

A new fermentation process that converts valueless organic waste streams into hydrogen-rich gas has been developed by Van Ginkel *et al.* (2001). The process employs mixed microbial cultures readily available in nature, such as compost, anaerobic digester sludge, soil, etc., to convert organic wastes into hydrogen-rich gas. The biodegradation efficiencies of the pollutants were examined by changing the hydraulic retention time (HRT) as a main operating variable. An enriched culture of hydrogen-producing bacteria such as *Clostridia* was obtained by heat treatment, pH control, and HRT control of the treatment system. The biohydrogen fermentation technology could enhance the economic viability of many processes utilizing hydrogen as a fuel source or as raw material. Figure 5.9 shows the basic components of an anaerobic digestion system.

Anaerobic fermentative microorganisms, cyanobacteria, and algae are suitable in the biological production of hydrogen via hydrogenase due to reversible hydrogenases (Adams 1990). Cyanobacteria and algae can carry out the photoevolution of hydrogen catalyzed by hydrogenases. The reactions are similar to electrolysis involving splitting of water into oxygen and hydrogen (Gaffron 1940).

Biological hydrogen can be generated from plants by biophotolysis of water using microalgae (green algae and cyanobacteria), fermentation of organic compounds, and photodecomposition of organic compounds by photosynthetic bacteria. To produce hydrogen by fermentation of biomass, a continuous process using a nonsterile substrate with a readily available mixed microflora is needed (Hussy *et al.* 2005). A successful biological conversion of biomass into hydrogen depends strongly on the processing of raw materials to produce feedstock, which can be fermented by the microorganisms.

5.3 Liquefaction of Algal Cells

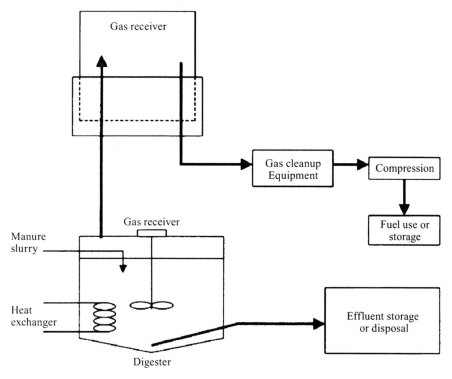

Figure 5.9 Basic components of an anaerobic digestion system

Hydrogen-producing bacteria (*Clostridia*) were found to have growth rates about 5 to 10 times higher than that of methane-producing bacteria (Van Ginkel et al. 2001). In a continuous-flow bioreactor system, hydrogen production showed a declining trend in the later stages of reactor operation. Based on these findings, it is hypothesized that *Clostridia* may have gone through a phenomenon known as "degeneration" in which they lose their ability to produce hydrogen. Therefore, inoculating fresh mixed cultures may be a feasible way to maintain sustainable hydrogen production. Based on this hypothesis, a two-stage anaerobic reactor has been proposed. The first-stage reactor is designed as a hydrogen-producing reactor, whereas the second-stage reactor will be employed to cultivate fresh seed culture to perpetually supply to the first one.

5.3 Liquefaction of Algal Cells

Hydrocarbons of algal cells have been separated by extraction with organic solvent after freeze-drying and sonicating the algal cells. However, these procedures are not suitable for separation on a large scale because they are costly. Therefore, an effective method is liquefaction for separating hydrocarbons as liquid fuel from harvested algal cells with high moisture content. The direct thermochemical liquefaction can convert wet biomass such as wood and sewage sludge into liquid fuel at around 575 K and 10 MPa using a catalyst such as sodium carbonate (Demirbas 2007). At the same time, the liquid oil can be easily separated (Ogi *et al.* 1990).

Processes relating to liquefaction of biomass are based on the early research of Appell *et al.* (1971). These workers reported that a variety of biomass such as agricultural and municipal wastes could be converted, partially, into a heavy oil-like product by reaction with water and carbon monoxide/hydrogen in the presence of sodium carbonate. The heavy oil obtained from the liquefaction process is a viscous tarry lump, which sometimes causes troubles in handling. For this purpose, some organic solvents can be added to the reaction system. These processes require high temperature and pressure.

In the liquefaction process, biomass is converted into liquefied products through a complex sequence of physical structure and chemical changes. The feedstock of liquefaction is usually a wet matter. In liquefaction, biomass is decomposed into small molecules. These small molecules are unstable and reactive and can repolymerize into oily compounds with a wide range of molecular weight distributions (Demirbas 2000).

Liquefaction can be accomplished directly or indirectly. Direct liquefaction involves rapid pyrolysis to produce liquid tars and oils or condensable organic vapors. Indirect liquefaction involves the use of catalysts to convert noncondensable, gaseous products of pyrolysis or gasification into liquid products. Alkali salts, such as sodium carbonate and potassium carbonate, can induce the hydrolysis of cellulose and hemicellulose into smaller fragments. The degradation of biomass into smaller products mainly proceeds by depolymerization and deoxygenation. In the liquefaction process, the amount of solid residue increases in proportion to the lignin content. Lignin is a macromolecule, which consists of alkylphenols and has a complex three-dimensional structure. It is generally accepted that free phenoxyl radicals are formed by thermal decomposition of lignin above 500 K and that the radicals have a random tendency to form a solid residue through condensation or repolymerization (Demirbas 2000).

The changes during liquefaction process involve all kinds of processes such as solvolysis, depolymerization, decarboxylation, hydrogenolysis, and hydrogenation. Solvolysis results in micellarlike substructures of the biomass. The depolymerization of biomass leads to smaller molecules. It also leads to new molecular rearrangements through dehydration and decarboxylation. When hydrogen is present, hydrogenolysis and hydrogenation of functional groups, such as hydroxyl groups, carboxyl groups, and keto groups, also occur.

Direct hydrothermal liquefaction in subcritical water conditions is a technology that can be employed to convert wet biomass material into liquid fuel. A number of technical terminologies have been used in the literature to refer to this technology, but it essentially utilizes the high activity of water in subcritical conditions in order to decompose biomass materials down into shorter and smaller molecular materials with a higher energy density or more valuable chemicals.

5.3 Liquefaction of Algal Cells 129

Past research in the use of hydrothermal technology for direct liquefaction of algal biomass was very active. Minowa *et al.* (1995) reported an oil yield of about 37% (organic basis) by direct hydrothermal liquefaction at around 300 °C and 10 MPa from *Dunaliella tertiolecta* with a moisture content of 78.4%wt. The oil obtained at a reaction temperature of 340 °C and holding time of 60 min had a viscosity of 150 to 330 mPas and a calorific value of 36 kJ/g, comparable to those of fuel oil. It was concluded that the liquefaction technique was a net energy producer from the energy balance. In a similar study on oil recovery from *Botryococcus braunii*, a maximum yield 64% dry wt. basis of oil was obtained by liquefaction at 300 °C catalyzed by sodium carbonate (Sawayama *et al.* 1995). Also, Aresta *et al.* (2005) have compared different conversion techniques, *viz.*, supercritical CO_2, organic solvent extraction, pyrolysis, and hydrothermal technology, for the production of microalgal biodiesel. The hydrothermal liquefaction technique was more effective for extracting microalgal biodiesel than supercritical CO_2. From these two studies, it is reasonable to believe that, among the selected techniques, hydrothermal liquefaction is the most effective technological option for the production of biodiesel from algae. Nevertheless, due to the level of limited information in the hydrothermal liquefaction of algae, more research in this area is needed.

Liquefaction of *B. braunii*, a colony-forming microalga, with high moisture content was performed with or without sodium carbonate as a catalyst for conversion into liquid fuel and recovery of hydrocarbons. A greater amount of oil than the content of hydrocarbons in *B. braunii* (50 wt% db) was obtained, in a yield of 57 to 64 wt% at 575 K. The oil was equivalent in quality to petroleum oil. The recovery of hydrocarbons was maximized (>95%) at 575 K (Banerjee *et al.* 2002).

5.3.1 Liquefaction of Algal Cells by Hexane Extraction

Hexane solubles of raw algal cells are shown in Table 5.14. Hexane soluble was obtained at a high yield of 58% of its dry weight and had good fluidity with a viscosity of 56 cP and a high heating value (49 MJ/kg). The properties of primary oil are shown in Table 5.15. The yield of the primary oil obtained at 575 K was 52.9% and that at 475 K was 56.5%; these values were slightly lower than the yield of the hexane soluble of the raw algal cells. This suggests that hydrocarbons of raw algal cells were partly converted into dichloromethane insoluble materials such as char (Dote *et al.* 1994).

The heating value of the primary oil obtained at 575 K was 47.5 MJ/kg and that at 475 K was 42.0 MJ/kg; these values were equivalent to petroleum oil. In particular, the heating value of the primary oil obtained at 575 K was much higher than that of the oil obtained by liquefaction of other biomasses. The viscosity of the primary oil obtained at 575 K was as low (94 cP) as that of the hexane soluble of the raw algal cells. However, the viscosity of the primary oil obtained at 475 K was too high to measure: the primary oil was like rubber. Therefore, the primary oil obtained at 575 K could be used as fuel oil. The oxygen content of the primary oil obtained

130

5 Energy from Algae

Table 5.14 Properties of microalgae used for liquefaction

Moisture content	Dry solid	Ash[a]	Organics[a]	Elemental analysis (%)[a]			
(%)	(%)	(%)	(%)	C	H	N	O
92.0	8.0	2.0	98.0	68.7	10.9	1.3	19.1

[a] On a dry algal cells basis.

Table 5.15 Properties of the hexane soluble of raw algal cells

Yield	Heating value	Viscosity	Elemental analysis (%)			
(%)[a]	(MJ/kg)	(cP, at 50 °C)	C	H	N	O
58.0	49.4	56.0	84.6	14.5	0.1	0.9

[a] On a dry-algal-cell basis.

at 575 K was slightly higher than that of the hexane soluble of the raw algal cells. However, it was much lower than that of the oil obtained by liquefaction of other biomasses (Dote *et al.* 1994).

The properties of the hexane soluble of primary oil are shown in Table 5.16. The yield of the hexane soluble of the primary oil obtained at 575 K was 44% and that at 475 K was 39% on a dry-algal-cell basis. This means that the primary oil obtained at 575 K contained 83% of hexane soluble and that at 475 K contained 69% of hexane soluble. The elemental composition of the three hexane solubles was almost equal. The hexane solubles of the primary oil obtained at 575 and 475 K had good fluidity, as did the hexane soluble of the raw algal cells. Despite thermal treatment at high temperature, the hexane soluble of primary oil has properties that are similar to those of the hexane soluble of raw algal cells (Dote *et al.* 1994). Figure 5.10 shows the liquefaction of algal cells by hexane extraction obtained for primary oil.

Table 5.16 Properties of hexane soluble of primary oil

Temperature	Yield	Heating value	Viscosity	Elemental analysis (%)[a]			
(°C)	(%)[a]	(MJ/kg)	(cP, at 50 °C)	C	H	N	O
300	44.0	49.2	77	84.5	14.3	0.1	1.1
200	39.0	50.3	170	83.5	15.3	0.1	1.1

[a] On a dry algal cells basis.

5.4 High-value-added Products from Algae

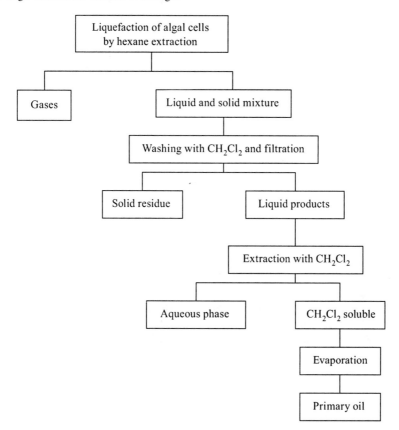

Figure 5.10 Liquefaction of algal cells by hexane extraction obtained for primary oil

5.4 High-value-added Products from Algae

5.4.1 Small Molecules

The microalga *Dunaliella salina* can contain up to 40% of its dry weight as glycerol. However, the low price of glycerol (as a coproduct of biodiesel production) means that the algal product would not be competitive. Other algal species accumulate high concentrations of proline under conditions of high salinity (Benemann and Oswald 1996).

Although a number of algal strains were investigated for growth and lipid production properties, the best candidates were found in two classes, *Chlorophyceae* (green algae) and *Bacilliarophyceae* (diatoms). Organisms were identified in both classes that showed high productivity, an ability to grow in large-scale culture, and lipid accumulation upon nutrient stress. However, in some ways the diatoms may turn out to be better candidate organisms for biofuel production.

132 5 Energy from Algae

A dozen potential by- and coproducts can be obtained from algae. These are iodine, algin, mannitol, and lignin-related fraction, the first three being commercial products. The lignin-related fraction was suggested as a feedstock or as a component for making specialty plastics, adhesives, and timed-release substances such as pharmaceuticals or pesticides (Chynoweth *et al.* 1993).

5.4.2 Polymers

Bioflocculation occurs in high-rate ponds. This process involves removing algae from the paddlewheel-mixed ponds and placing them in a quiescent container, where they would spontaneously flocculate and rapidly settle. There are several apparently distinct mechanisms by which algae flocculate and then settle, including "autoflocculation," which is induced by high pH in the presence of phosphate and divalent cations (Mg^{+2} and Ca^{+2}), and flocculation induced by N limitation. Bioflocculation refers to the tendency of normally repulsive microalgae to aggregate in large flocs, which then exhibit a rather high sedimentation velocity. The mechanisms of bioflocculation involve extracellular polymers excreted by algae, but the details remain to be investigated (Sheehan *et al.* 1998).

Organic flocculants at about 2 to 6 g/kg and $FeCl_3$ at about 15 to 200 g/kg of algal biomass are required to remove 90% or more of the algal cells. Because of the high cost of the organic flocculants, costs are comparable for both flocculants tested. The polymers can be used in very small amounts without contributing a major cost to the overall process (Sheehan *et al.* 1998).

Macroalgae have long been used for the production of phycocolloids such as alginates, carrageenans, or agars. They make up the major industrial products derived from algae (Radmer 1996; Pulz and Gross 2004). These polymers are located either in cell walls or within the cells serving as storage materials. A characteristic of marine algae is the abundance of sulfated polysaccharides in their cell walls.

5.4.3 High-value Oils

The very long-chain polyunsaturated fatty acids (vlcPUFAs) eicosapentaenoic (EPA), docosahexaenoic acid (DHA), and arachidonic acid (AA) are well known for their nutritional importance. As they confer flexibility, fluidity, and selective permeability properties to cellular membranes, they have been shown to be vital for brain development and beneficial for the cardiovascular system and for other important nutraceutical and pharmaceutical targets in human and animal health (Funk 2001; de Urquiza 2000; Colquhoun 2001). For example, vlcPUFAs are found in many different product applications including formulas for infants, adult dietary supplements, animal feed, food additives, and pharmaceutical precursors. These applications represent an extensive market for vlcPUFAs: the world wholesale market for infant formula alone is estimated to be about $ 10 billion per annum (Ward and Singh 2005).

Animals lack the capability to synthesize vlcPUFAs, and therefore these essential fatty acids must be obtained from food/feed. Typically sources of PUFAs are oil-rich fish such as eel, mackerel, herring, salmon, and sardines (Ward and Singh 2005). Due to concerns over declining fish stocks and the potential for fish oils to be contaminated by a range of pollutants, the possibilities for obtaining these fatty acids from other sources have been investigated (Qi *et al.* 2004). Interestingly, the vlcPUFAs in oil-rich fish originate from marine microalgae that are eaten by the fish. Algal genes encoding relevant enzymes have been identified, and recently several groups have reported progress on using these genes to produce DHA and ARA in transgenic plants, including crops such as soybean, linseed, tobacco, and the model species *Arabidopsis* (Qi *et al.* 2004; Abbadi *et al.* 2004). By adding additional genes to those that are needed to produce ARA and EPA, production of DHA has been established in soybean, *Brassica juncea*, and *Arabidopsis* (Robert *et al.* 2005; Wu *et al.* 2005).

An alternative approach is to use directly the algae that are the most efficient primary producers of the vlcPUFAs. Algae groups that contain vlcPUFAs include diatoms, crysophytes, cryptophytes, and dinoflagellates (Cohen *et al.* 1995; Behrens and Kyle 1996). High amounts of DHA, for example, are produced in the algae *Crypthecodinium cohnii, Thraustochytrium spp., Schizochytrium spp., Isochrysis galbana,* and *Crypthecodinium spp.* (Ward and Singh 2005). The algae *Porphyridium cruentum* and *Parietochloris incise* accumulate AA (Zhang *et al.* 2002; Guil-Guerrero *et al.* 2000) and several species have been suggested for the production of EPA including *Nitzschia spp., Nannochloropsis spp., Navicula spp., Phaeodatylum spp.,* and *Porphyridium spp.* (Tan and Johns 1996; Sukenik 1991; Molina Grima *et al.* 2003; Cohen *et al.* 1995). For additional information about the content of vlcPUFAs in different microalgae see Barclay *et al.* (1994), Wen and Chen (2003), and Ward and Singh (2005). A slight inconvenience with using algal feedstocks directly for the production of vlcPUFAs is that in many species the accumulation of these fatty acids involves their presence in lipids other than triacylglycerides such as galactolipids. This makes their isolation more complicated. For vlcPUFA production directly from microalgae it has been estimated that the cost of producing EPA from *Phaeodactylum tricornutum* cultured in photobioreactors is about $ 4602 kg-1 (Molina Grima *et al.* 2003).

References

Abbadi, A., Domergue, F., Bauer, J., Napier, J. A., Welti, R., Ohringer, U., Cirpus, P., Heinz, E. 2004. Biosynthesis of very-long-chain polyunsaturated fatty acids in transgenicoilseeds: constraints on their accumulation. Plant Cell 16:2734–2746.

Ackman, R. G., Tocher, C. S., McLachlan, J. 1968. Marine phytoplankter fatty acids. J Fish Res Board Can 25:1603–1620.

Adams, M. W. W. 1990. The structure and mechanism of iron-hydrogenases. Biochim Biophys Acta 1020:115–145.

Appel, H. R., Fu, Y. C., Friedman, S., Yavorsky, P. M, Wender, I. 1971. Converting organic wastes to oil. US Burea of Mines Report of Investigation, No. 7560, Washington, D.C.

Aresta, M., Dibenedetto, A., Barberio, G. 2005a. Utilization of macro-algae for enhanced CO_2 fixationand biofuels production: Development of a computing software for an LCA study. Fuel Process Technol 86:1679–1693.

Bala, B. K. 2005. Studies on biodiesels from transformation of vegetable oils for diesel engines. Energy Educ Sci Technol 15:1–45.

Banerjee, A., Harma, R.S., Chisti, Y., Banerjee, U. C. 2002. *Botryococcus braunii*: a renewable source of hydrocarbons and other chemicals. Crit Rev Biotechnol 22:245–279.

Barclay, W. R., Meager, K. M., Abril, J. R. 1994. Heterotrophic production of long chain omega-3 fatty acids utilizing algae and algae-like microorganisms. J Appl Phycol 6:123–129.

Behrens, P. W., Kyle, D. J. 1996. Microalgae as a source of fatty acids. J Food Lipids 3:259–272.

Bender, M. 1999. Economic feasibility review for community-scale farmer cooperatives for biodiesel. Biores Technol 70:81–87.

Benemann, J., Oswald, W. J. 1996. Systems and economic analysis of microalgae ponds for conversion of CO_2 to biomass. US DOE, Pittsburgh Energy Technology Centre, Pittsburgh, PA.

Benemann, J. R., Pursoff, P., Oswald, W. J. 1978. Engineering Design and Cost Analysis of a Large-Scale Microalgae Biomass System, NTIS#H CP/T1605-01 UC-61). US Department of Energy, Washington, D.C.

Burlew, J. S. 1953. Algal Culture: From Laboratory to Pilot Plant (Publication No. 600). Carnegie Institution of Washington, Washington, D.C.

Caliceti, M., Argese, E., Sfriso, A., Pavoni, B. 2002. Heavy metal contamination in the seaweeds of the Venice lagoon. Chemosphere 47:443–454.

Cardone, M., Prati, M. V., Rocco, V., Senatore, A. 1998. Experimental analysis of performances and emissions of a diesel engines fuelled with biodiesel and diesel oil blends. Proceedings of MIS–MAC V, Rome, Italy, pp. 211–25 (in Italian).

Carraretto, C., Macor, A., Mirandola, A., Stoppato, A., Tonon, S. 2004. Biodiesel as alternative fuel: experimental analysis and energetic evaluations. Energy 29:2195–2211.

Chisti, Y. 2007. Biodiesel from microalgae. Biotechnol Adv 25:294–306.

Chynoweth, D. P., Turick, C. E., Owens, J. M., Jerger, D. E., Peck, M. W. 1993. Biochemical methane potential of biomass and waste feedstocks. Biomass Bioenergy 5:95–111.

Cohen, Z., Norman, H. A., Heimer, Y. M. 1995. Microalgae as a source of omega 3 fatty acids. World Rev Nutrit Dietet 77:1–31.

Colquhoun, D. M. 2001. Nutraceuticals: vitamins and other nutrients in coronary heart disease. Current Opin Lipidol 12:639–646.

Collyer, D. M., Fogg, G. E. 1955. Studies of fat accumulation by algae. J Exp Bot 6:256–275.

Constantopolous, G., Bloch, K. 1967. Effect of light intensity on the lipid composition of Euglena gracilis. J Biol Chem 242:3538–3542.

Coombs, J., Darley, W. M., Holm-Hansen, O., Volcani, B. E. 1967. Studies on the biochemistry and fine structure of silica shell formation in diatoms. Chemical composition of Navicula pelliculosa during silicon starvation. Plant Physiol 42:1601–1606.

Coronado, C. R., de Carvalho Jr., J. A., Silveira, J. L. 2009. Biodiesel CO_2 emissions: a comparison with the main fuels in the Brazilian market. Fuel Proc Technol 90:204–211.

Corro, G., Ayala, E. 2008. Bioethanol and diesel/bioethanol blends emissions abatement. Fuel 87:3537–3542.

Dene, T., Hole, J. 2006. Enabling biofuels: biofuel economics. Final Report. Minister of Transport, Covec Ltd., Auckland, New Zealand, June 2006.

Demirbas, A. 2000. Mechanisms of liquefaction and pyrolysis reactions of biomass. Energy Convers Manage 41:633–646.

Demirbas, A. 2002. Biodiesel from vegetable oils via transesterification in supercritical methanol. Energy Convers Manage 43:2349–2356.

References 135

Demirbas, A. 2003. Biodiesel fuels from vegetable oils via catalytic and non-catalytic supercritical alcohol transesterifications and other methods: a survey. Energy Convers Manage 44:2093–2109.

Demirbas, A. 2006. Oily products from mosses and algae via pyrolysis. Energy Sources A 28:933–940.

Demirbas, A. 2007. Production of biofuels from macroalgae and microalgae. Energy Educ Sci Technol A 18:59–65.

Demirbas, A. 2008. Economic and environmental impacts of the liquid biofuels. Energy Educ Sci Technol A 22:37–58.

Demirbas, A. H. 2009a. Inexpensive oil and fats feedstocks for production of biodiesel. Energy Educ Sci Technol A 23:1–13.

Demirbas, A. 2009b. Political, economic and environmental impacts of biofuels: a review. Appl Energy 86:2627–2636.

Demirbas, A., Arin, G. 2004. Hydrogen from biomass via pyrolysis: relationships between yield of hydrogen and temperature. Energy Sources 26:1061–1069.

De Urquiza, A. M., Liu, S., Sjoberg, M., Zetterstrom, R. H., Griffiths, W., Sjovall, J., Perlmann, T. 2000. Docosahexaenoic acid, a ligand for the retinoid X receptor in mouse brain. Science 290:2140–2144.

Dorado, M. P., Ballesteros, E. A., Arnal, J. M., Gomez, J., Lopez, F. J. 2003. Exhaust emissions from a Diesel engine fueled with transesterified waste olive oil. Fuel 82:1311–1315.

Dote, Y., Sawayama, S., Inoue, S., Minowa, T., Yokoyama, S. 1994. Recovery of liquid fuel from hydrocarbon-rich microalgae by thermochemical liquefaction. Fuel 73:1855–1857.

Dufey, A. 2006. Biofuels production, trade and sustainable development: emerging issues. Environmental Economics Programme, Sustainable Markets Discussion Paper No. 2, International Institute for Environment and Development (IIED), London, September 2006.

Dunn, R. O. 2001. Alternative jet fuels from vegetable-oils. Trans ASAE 44:1151–1757.

Dote, Y., Sawayama, S., Yokoyama, S. 1995. Liqefaction of hydrocarbon-rich microalga National Institute for Resources and Environment 16-3, Onogawa, Tsukuba, Ibaraki 305, Japan.

EPA. 2002. A comprehensive analysis of biodiesel impacts on exhaust emissions. Draft Technical Report, EPA420-P-02-001, US Environmental Protection Agency, October 2002.

Filipkowska, A., Lubecki, L., Szymczak-Żyła, M., Kowalewska, G., Żbikowski, R., Szefer, P. 2008. Utilisation of macroalgae from the Sopot beach (Baltic Sea). Oceanologia 50:255–273.

Funk, C. D. 2001. Prostaglandins and leukotrienes: advances in eicosanoid biology. Science 294:1871–1875.

Gaffron, H. 1940. Carbon dioxide reduction with molecular hydrogen in green algae. Am J Bot 27:273–283.

Gercel, H. F. 2002. The effect of a sweeping gas flow rate on the fast pyrolysis of biomass. Energy Sources 24:633–642.

Graham, L. A., Belisle, S. L., Baas, C. L. 2008. Emissions from light duty gasoline vehicles operating on low blend ethanol gasoline and E85. Atmos Environ 42:4498–4516.

Granda, C. B., Zhu, L., Holtzapple, M. T. 2007. Sustainable liquid biofuels and their environmental impact. Environ Prog 26:233–250.

Grassi, G. 1999. Modern bioenergy in the European Union. Renew Energy 16:985–990.

Guil-Guerrero, J. L., Belarbi, E. H., Rebolloso-Fuentes, M. M. 2000. Eicosapentaenoic and arachidonic acids purification from the red microalga Porphyridium cruentum. Bioseparation 9:299–306.

Gunaseelan, V. N. 1997. Anaerobic digestion of biomass for methane production: a review. Biomass Bioenergy 13:83–114.

Haas, M. J., McAloon, A. J., Yee, W. J., Foglia, T. A. 2006. A process model to estimate biodiesel production costs. Bioresour Technol 97:671–678.

Hussy, I., Hawkes, F. R., Dinsdale, R., Hawkes, D. L. 2005. Continuous fermentative hydrogen production from sucrose and sugarbeet. Int J Hydrogen Energy 30:471–483.

IEA. 2007. Biodiesel statistics. IEA Energy Technology Essentials, OECD/IEA. International Energy Agency, Paris, January 2007.

Iwasa, N., Kudo, S., Takahashi, H., Masuda, S., Takezawa, N. 1993. Highly selective supported Pd catalysts for steam reforming of methanol. Catal Lett 19:211–216.

Janulis, P. 2004. Reduction of energy consumption in biodiesel fuel life cycle. Renew Energy 29:861–871.

Jenkins, B. M., Baxter, L. L., Miles Jr., T. R., Miles, T. R. 1998. Combustion properties of biomass. Fuel Proc Technol 54:17–46.

Kim, H., Choi, B. 2008. Effect of ethanol-diesel blend fuels on emission and particle size distribution in a common-rail direct injection engine with warm-up catalytic converter. Renew Energy 33:2222–2228.

Kishimoto, M., Okakura, T., Nagashima, H., Minowa, T., Yokoyama, S., Yamaberi, K., 1994. CO_2 fixation and oil production using microalgae. J Ferment Bioeng 78:479–482.

Krahl, J., Knothe, G., Munack, A., Ruschel, Y., Schröder, O., Hallier, E., Westphal, G., Bünger, J. 2009. Comparison of exhaust emissions and their mutagenicity from the combustion of biodiesel, vegetable oil, gas-to-liquid and petrodiesel fuels. Fuel 88:1064–1069.

Laforgia, D., Ardito, V. 1994. Biodiesel fuelled IDI engines: performances, emissions and heat release investigation. Biores Technol 51:53–59.

Lapuerta, M., Armas, O., Herreros, J. M. 2008. Emissions from a diesel-bioethanol blend in an automotive diesel engine. Fuel 87:25–31.

Li, Y., Horsman, M., Wu, N., Lan, C. Q., Dubois-Calero, N. 2008. Biofuels from microalgae. Biotechnol Prog 24:815–820.

Meier, R. L. 1955. Biological cycles in the transformation of solar energy into useful fuels. In: Daniels, F., Duffie, J.A. (eds.). Solar Energy Research. University of Wisconsin Press, Madison, WI, pp. 179–183.

Miao, X., Wu, Q. 2004. High yield bio-oil production from fast pyrolysis by metabolic controlling of Chlorella protothecoides. J Biotechnol 110:85–93.

Minowa, T., Yokoyama, S.-Y., Kishimoto, M., Okakura, T. 1995. Oil production from algal cells of dunaliella tertiolecta by direct thermochemical liquefaction. Fuel 74:1735–1738.

Miyake, J., Veziroglu, T. N., Takashashi, P. K. 1990. Hydrogen energy progress VIII. Proceedings of the 8th WHEC, Hawaii, Pergamon, New York.

Molina Grima, E. M., Belarbi, E. H., Fernandez, F. G. A., Medina, A. R., Chisti, Y. 2003. Recovery of microalgal biomass and metabolites: process options and economics. Biotechnol Adv 20:491–515.

Najafi, G., Ghobadian, B., Tavakoli, T., Buttsworth, D. R., Yusaf, T. F., Faizollahnejad, M. 2009. Performance and exhaust emissions of a gasoline engine with ethanol blended gasoline fuels using artificial neural network. Appl Energy 86:630–639.

Nath, K., Das, D. 2003. Hydrogen from biomass. Curr Sci 85:265–271.

Nichols, B. W. 1965. Light induced changes in the lipids of Chlorella vulgaris. Biochim Biophys Acta 106:274–279.

Nithedpattrapong, S., Srikul, S., Korawis, C., Onthong, J. 1995. Influence of N, P, K and Mg on yield of oil palm grown on Kohong soil series. Thai Agricult Res J 13:164–174.

Ogi, T., Yokoyama, S., Minowa,T., Dote, Y. 1990. Role of butanol solvent in direct liquefaction of wood. Sekiyu Gakkaishi (J Jpn Petrol Inst) 33:383–389.

Oswald, W.J., Golueke, C. 1960. Biological transformation of solar energy. Adv Appl Microbiol 2:223–262.

Peng, W. M., Wu, Q. Y., Tu, P. G. 2000. Effects of temperature and holding time on production of renewable fuels from pyrolysis of Chlorella protothecoides. J Appl Phycol 12:147–152.

Peng, W. M., Wu, Q. Y., Tu, P. G. 2001. Pyrolytic characteristics of heterotrophic Chlorella protothecoides for renewable bio-fuel production. J Appl Phycol 13:5–12.

Pohl, P., Wagner, H. 1972. Control of fatty acid and lipid biosynthesis in Euglena gracilis by ammonia, light and DCMU. Z Naturforsch 27:53–61.

References

Pulz, O., Gross, W. 2004. Valuable products from biotechnology of microalgae. Appl Microbiol Biotechnol 65:635–648.

Putun, A. E. 2002. Biomass to bio-oil via fast pyrolysis of cotton straw and stalk. Energy Sour 24:275–285.

Qi, B., Fraser, T., Mugford, S., Dobson, G., Sayanova, O., Butler, J., Napier, J. A., Stobart, A. K., Lazarus, C. M. 2004. Production of very long chain polyunsaturated omega-3 and omega-6 fatty acids in plants. Nat Biotechnol 22:739–745.

Radmer, R. J. 1996. Algal diversity and commercial algal products. Bioscience 46:263–270.

Rakopoulos, D. C., Rakopoulos, C. D., Kakaras, E. C., Giakoumis, E. G. 2008. Effects of ethanol-diesel blends on the performance and exhaust emissions of heavy duty DI diesel engine. Energy Convers Manage 49:3155–3162.

RFA. 2007. Ethanol Industry Statistics. Renewable Fuels Association, Washington, D.C.

RFA. 2009. Ethanol Industry Statistics. Renewable Fuels Association, Washington, D.C.

Robert, S. S., Singh, S. P., Zhou, X. R., Petrie, J. R., Blackburn, S. I., Mansour, M. P., Nichols, P. D., Liu, Q., Green, A. G. 2005. Metabolic engineering of Arabidopsis to produce nutritionally important DHA in seed oil. Funct Plant Biol 32:473–479.

Rosenberg, A., Gouaux, J. 1967. Quantitative and compositional changes in monogalactosyl and digalactosyl diglycerides during light-induced formation of chloroplasts in Euglena gracilis. J Lipid Res 8:80–83.

Sawayama, S., Inoue, S., Dote, Y., Yokoyama, S.-Y. 1995. CO_2 fixation and oil production through microalga. Energy Convers Manage 36:729–731.

Sawayama, S., Minowa, T., Yokoyama, S. Y. 1999. Possibility of renewable energy production and CO_2 mitigation by thermochemical liquefaction of microalgae. Biomass Bioenergy 17:33–39.

Shay, E. G. 1993. Diesel fuel from vegetable oils: status and opportunities. Biomass Bioenergy 4:227–242.

Sheehan, J., Dunahay, T., Benemann, J., Roessler, P. 1998. A Look Back at the U.S. Department of Energy's Aquatic Species Program – Biodiesel from Algae. National Renewable Energy Laboratory (NREL) Report: NREL/TP-580-24190. Golden, CO.

Soeder, C.J. 1986. A historical outline of applied algology. In: Richmond, A. (ed.). Handbook of Microalgal Mass Culture. CRC, Boca Raton, FL, pp. 25–41.

Soltic, P., Edenhauser, D., Thurnheer, T., Schreiber, D., Sankowski, A. 2009. Experimental investigation of mineral diesel fuel, GTL fuel, RME and neat soybean and rapeseed oil combustion in a heavy duty on-road engine with exhaust gas after treatment. Fuel 88:1–8.

Spoehr, H. A., Milner, H. W. 1949. The chemical composition of Chlorella: effect of environmental conditions. Plant Physiol 24:120–149.

Sukenik, A. 1991. Ecophysiological considerations in the optimization of eicosapentaenoic acid production by Nannochloropsis sp. (Eustigmatophyceae). Biores Technol 35:263–269.

Takezawa, N., Shimokawabe, M., Hiramatsu, H., Sugiura, H., Asakawa, T., Kobayashi, H. 1987. Steam reforming of methanol over Cu/ZrO_2. Role of ZrO_2 support. React Kinet Catal Lett 33:191–196.

Tan, C. K., Johns, M. R. 1996. Screening of diatoms for heterotrophic eicosapentaenoic acid production. J Appl Phycol 8:59–64.

Thamsiriroj, T. 2007. Optimal biomass and technology for production of biofuel as a transport fuel. Master's thesis, University College, Cork, UK.

Thamsiriroj, T., Murphy, J. D. 2009. Is it better to import palm oil from Thailand to produce biodiesel in Ireland than to produce biodiesel from indigenous Irish rape seed? Appl Energy 86:595–604.

Tilman, D., Hill, J., Lehman, C. 2006. Carbon-negative biofuels from low-input high diversity grassland biomass. Science 314:1598–1600.

Tsukahara, K., Sawayama, S. 2005. Liquid fuel production using microalgae. J Jpn Petr Inst 48:251–259.

138 5 Energy from Algae

Tzirakis, E., Karavalakis, G., Zannikos, F., Stournas, S. 2007. Impact of Diesel/biodiesel blends on emissions from a diesel vehicle operated in real driving conditions. SAE Technical Paper, National Technical University of Athens, Athens, Greece.

UN. 2006. The emerging biofuels market: regulatory, trade and development implications. United Nations Conference on Trade and Development, New York and Geneva.

Urbanchuk, J. M. 2007. Economic impacts on the farm community of cooperative ownership of ethanol production, LECG, LLC, Wayne, PA, 13 February 2007.

Van Ginkel, S., Sung, S., Lay, J. J. 2001. Biohydrogen production as a function of pH and substrate concentration. Environ Sci Technol 35:4726–4730.

Ward, O. P., Singh, A. 2005. Omega-3/6 fatty acids: alternative sources of production. Proc Biochem 40:3627–3652.

Wen, Z. Y., Chen, F. 2003. Heterotrophic production of eicosapentaenoic acid by microalgae. Biotechnol Adv 21:273–294.

Werner, D. 1966. Die Kieselsaure im Stoffwechsel von Cyclotella cryptica Reimann, Lewin and Guilard. Arch Mikrobiol 55:278–308.

Whims, J. 2002. Corn based ethanol costs and margins. Attachment 1, AGMRC, Kansas State Uuniversity.

Wu, G. H., Truksa, M., Datla, N., Vrinten, P., Bauer, J., Zank, T., Cirpus, P., Heinz, E., Qiu, X. 2005. Stepwise engineering to produce high yields of very long-chain polyunsaturated fatty acids in plants. Nat Biotechnol 23:1013–1017.

Xu, H., Miao, X., Wu, Q. 2006. High quality biodiesel production from a microalga Chlorella prototothecoides by heterotrophic growth in fermenters. J Biotechnol 126:499–507.

Yu, S., Tao, J. 2009. Simulation based life cycle assessment of airborne emissions of biomass-based ethanol products from different feedstocks planting areas in China. J Cleaner Prod 17:501–506.

Zhang, C. W., Cohen, Z., Khozin-Goldberg, I., Richmond, A. 2002. Characterization of growth and arachidonic acid production of Parietochloris incisa comb. nov (Trebouxiophyceae, Chlorophyta). J Appl Phycol 14:453–460.

Zhang, Y., Dub, M. A., McLean, D. D., Kates, M. 2003. Biodiesel production from waste cooking oil: 2. Economic assessment and sensitivity analysis. Biores Technol 90:229–240.

Chapter 6
Biodiesel from Algae

6.1 Introduction

Continued use of petroleum sourced fuels is now widely recognized as unsustainable because of depleting supplies and the contribution of these fuels to the accumulation of carbon dioxide in the environment. Renewable, carbon-neutral transport fuels are necessary for environmental and economic sustainability (Chisti 2007). Biodiesel can be carbon neutral and produced intensively on relatively small areas of marginal land. The quality of the fuel product is comparable to petroleum diesel and can be incorporated with minimal change into the existing fuel infrastructure. Innovative techniques, including the use of industrial and domestic waste as fertilizer, could be applied to further increase biodiesel productivity (Campbell 2008).

Algae, like corn, soybeans, sugar cane, wood, and other plants, use photosynthesis to convert solar energy into chemical energy. They store this energy in the form of oils, carbohydrates, and proteins. The plant oil can be converted into biodiesel; hence biodiesel is a form of solar energy. The more efficient a particular plant is at converting that solar energy into chemical energy, the better it is from a biodiesel perspective, and algae are among the most photosynthetically efficient plants on earth.

Algae can be a replacement for oil-based fuels, one that is more effective and has no disadvantages. Algae are among the fastest growing plants in the world, and about 50% of their weight is oil. This lipid oil can be used to make biodiesel for cars, trucks, and airplanes. Microalgae have much faster growth rates than terrestrial crops. The per-unit area yield of oil from algae is estimated to be between 20,000 and 80,000 L/acre/year; this is 7 to 31 times greater than the next best crop, palm oil. The lipid and fatty acid contents of microalgae vary in accordance with culture conditions. Most current research on oil extraction is focused on microalgae to produce biodiesel from algal-oil. Algal-oil can be processed into biodiesel as easily as oil derived from land-based crops.

The production of microalgal biodiesel requires large quantities of algal biomass. Macro- and microalgae are currently mainly used for food, in animal feed, in feed

A. Demirbas, M. Fatih Demirbas, *Algae Energy*   139
DOI 10.1007/978-1-84996-050-2, © Springer 2010

for aquaculture, and as biofertilizer. A 1-ha algae farm on wasteland can produce over 10 to 100 times as much oil compared to any other known source of oil crops. While a crop cycle may take from 3 months to 3 years for production, algae can start producing oil within 3 to 5 d, and thereafter oil can be harvested on a daily basis (just like milk). Algae can be grown using sea water and nonpotable water on wastelands where nothing else grows. Algae farming for biofuels is expected to provide a conclusive solution to the food vs. fuel debate.

The production of biodiesel has recently received much attention worldwide. In order to resolve the worldwide energy crisis, seeking for lipid-rich biological materials to produce biodiesel effectively has attracted much renewed interest. Algae have emerged as one of the most promising sources for biodiesel production. It can be inferred that algae grown in CO_2-enriched air can be converted into oily substances. Such an approach can contribute to solving the major problems of air pollution resulting from CO_2 emissions and future crises due to a shortage of energy sources (Sharif Hossain *et al.* 2008).

6.2 Biodiesel from Algal Oil

Biodiesel is defined as the monoalkyl esters of vegetable oils or animal fats. The vegetable oils and fats as alternative engine fuels are all extremely viscous, with viscosities ranging from 10 to 17 times greater than petroleum diesel fuel. Biodiesel is produced by transesterifying the parent oil or fat to achieve a viscosity close to that of petrodiesel. The chemical conversion of the oil into its corresponding fatty ester (biodiesel) is called transesterification. The purpose of the transesterification process is to lower the viscosity of the oil. The transesterification reaction proceeds with or without a catalyst by using primary or secondary monohydric aliphatic alcohols having one to four carbon atoms as follows (Demirbas 2007):

Triglycerides + Monohydric alcohol → Glycerin + Monoalkyl esters (Biodiesel)

(6.1)

This is an equilibrium reaction (Figure 6.1) where triglycerides can be processed into biodiesel, usually in the presence of a catalyst, and alkali such as potassium hydroxide (Ma and Hanna 1999; Demirbas 2007).

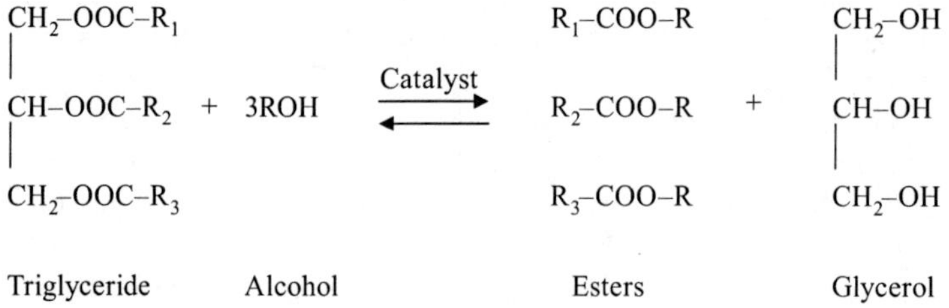

$$
\begin{array}{lllll}
CH_2\text{-OOC-R}_1 & & R_1\text{-COO-R} & & CH_2\text{-OH} \\
| & & & & | \\
CH\text{-OOC-R}_2 & + \ 3ROH \ \xrightleftharpoons{\text{Catalyst}} & R_2\text{-COO-R} & + & CH\text{-OH} \\
| & & & & | \\
CH_2\text{-OOC-R}_3 & & R_3\text{-COO-R} & & CH_2\text{-OH} \\
\\
\text{Triglyceride} & \text{Alcohol} & \text{Esters} & & \text{Glycerol}
\end{array}
$$

Figure 6.1 Transesterification of triglycerides with alcohol

6.2 Biodiesel from Algal Oil 141

Transesterification and catalytic cracking has usually been adopted to convert fat in the cell of microalgae into gasoline and diesel. This kind of method was limited by low temperature, and the outcome function was highly influenced by the fat content. What was more, the fat content in the microalgae had to be very high; otherwise good economic performance would be hard to achieve (Xu *et al.* 2006).

Biodiesel is a biofuel commonly consisting of methyl esters that are derived from organic oils, plant or animal, through the process of tranesterification. The biodiesel transesterification reaction is very simple:

$$\text{Triglyceride} + 3\,\text{Methanol} \xrightarrow{\text{Catalyst}} \text{Glycerine} + 3\,\text{Methyl esters (Biodiesel)}$$

$$(6.2)$$

An excess of methanol is used to force the reaction to favor the right side of the equation. The excess methanol is later recovered and reused.

The triglyceride is a complex molecule that plants and animals use for storing chemical energy; in more simple terms, it is fat. The process of making biodiesel occurs as follows. (1) Triglycerides, methanol, and catalyst are placed in a controlled reaction chamber to undergo transesterification. (2) The initial product is placed in a separator to remove the glycerine byproduct. (3) The excess methanol is recovered from the methyl esters through evaporation. (4) The final biodiesel is rinsed with water, neutralized, and dried (Xu *et al.* 2006). Unlike petroleum fuels, the relative simplicity of biodiesel manufacture makes its production scalable. Many existing vendors are small-time producers. Biodiesel is a somewhat "mature" fuel and was used as a diesel alternative in the early 20th century (Demirbas 2007). This has allowed biodiesel to attain a level of "grassroots" popularity among environmental advocates and visionaries.

The energy density of biodiesel is comparable to petroleum diesel. The higher heating value of petroleum diesel is 42.7 MJ/kg. Values for biodiesel vary depending on the source of biomass. Typically, biodiesel derived from seed oils, such as rapeseed or soybean, produces 39.5 MJ/kg, while biomass derived from algae yields 41 MJ/kg (Demirbas 1998; Rakopoulos *et al.* 2006; Xu *et al.* 2006). Although the lower-energy biodiesels based on seed oils are the most common, they have enough energy density to make them a viable alternative to petroleum diesel.

Biodiesel can be made from virtually any source of renewable oil. Typical sources include restaurant waste oil, animal fats, and vegetable oils. The supply of waste oil is very limited; however, it is a popular source for small-scale, independent producers. Large commercial producers often use vegetable oils, such as soybean, rapeseed, palm, and corn oils. Unfortunately, biodiesel derived from seed oil diverts from the food supply and the increasing competition for seed causes the oil, and resulting biodiesel, to become increasingly expensive (Campbell 2008).

The main advantages of biodiesel as diesel fuel its portability, ready availability, renewability, higher combustion efficiency, and lower sulfur and aromatic content (Demirbas 2007). Adopting biodiesel has a number of advantages. First, because the fuel is derived from biomass, it does not contribute to atmospheric CO_2 emissions. Second, biodiesel emissions are, except for NO_x, lower than petroleum diesel. Combustion of biodiesel alone provides over a 90% reduction in total unburned hydrocarbons, and a 75 to 90% reduction in polycyclic aromatic hydrocarbons (PAHs). Biodiesel further provides significantly greater reductions in particulates and carbon monoxide than petroleum diesel fuel. Biodiesel provides a slight increase or decrease in nitrogen oxides depending on engine family and testing procedures. Many studies on the performances and emissions of compression ignition engines, fueled with pure biodiesel and blends with diesel oil, have been performed and are reported in the literature (Laforgia and Ardito 1994; Cardone *et al.* 1998). Third, the infrastructure needed for biodiesel already exists. Biodiesel can be used in existing diesel engines blended with petroleum diesel, or it can be run unblended in engines with minor modifications (Crookes 2006; Rakopoulos *et al.* 2006; Bowman *et al.* 2006). Because biodiesel has twice the viscosity of petroleum diesel, its lubrication properties can actually improve engine life (Bowman *et al.* 2006). Fourth, biodiesel has low toxicity and is biodegradable (Aresta *et al.* 2005; Demirbas 2007). The biodegradabilities of several biodiesels in the aquatic environment show that all biodiesel fuels are readily biodegradable. After 28 d all biodiesel fuels are 77 to 89% biodegraded; diesel fuel is only 18% biodegraded in the same period (Zhang 1996). The enzymes responsible for the dehydrogenation/oxidation reactions that occur in the process of degradation recognize oxygen atoms and attack them immediately (Zhang *et al.* 1998). Fifth, like petroleum diesel, biodiesel has a more complete combustion than gasoline, giving a cleaner burn (Bowman *et al.* 2006). The oxygen content of biodiesel improves the combustion process and decreases its oxidation potential. The structural oxygen content of a fuel improves its combustion efficiency due to an increase in the homogeneity of oxygen with the fuel during combustion. Because of this, the combustion efficiency of biodiesel is higher than that of petrodiesel. A visual inspection of injector types would indicate no difference between biodiesel fuels when tested on petrodiesel. The overall injector coking is considerably low.

The major disadvantages of biodiesel are its higher viscosity, lower energy content, higher cloud point and pour point, higher nitrogen oxide (NO_x) emissions, lower engine speed and power, injector coking, engine compatibility, high price, and greater engine wear. Biodiesel is not without problems. First, it does produce increased NO_x emissions, relative to petroleum diesel, owing to the higher compression ratios typically used in biodiesel engines (Crookes 2006; Pradeep and Sharma 2007). Second, biodiesel does reduce the power output of a diesel engine compared to petroleum diesel, although this is only around 2% overall (Schneider 2006). Third, the production of biodiesel results in glycerine byproducts and wash wastewater. Fourth, the price of biodiesel is typically higher than that of petroleum diesel. Fifth, and most importantly, the biomass feedstocks for making biodiesel are diverted from other important uses, typically food production.

The algae that are used in biodiesel production are usually aquatic unicellular green algae. This type of algae is a photosynthetic eukaryote characterized by high growth rates and high population densities. Under good conditions, green algae can

6.2 Biodiesel from Algal Oil

double their biomass in less than 24 h (Schneider 2006; Chisti 2007). Additionally, green algae can have huge lipid contents, frequently over 50% (Schneider 2006; Chisti 2007). This high-yield, high-density biomass is ideal for intensive agriculture and may be an excellent source for biodiesel production.

The annual productivity and oil content of algae is far greater than that of seed crops. Soybean can only produce about 450 L of oil per hectare. Canola can produce 1,200 L/ha, and palm can produce 6,000 L. Now, compare that to algae, which can yield 90,000 L/ha (Haag 2007; Schneider 2006; Chisti 2007). It is possible that US demand for liquid fuel could be achieved by cultivating algae in one tenth the area currently devoted to soybean cultivation (Scott and Bryner 2006).

The process for producing microalgal oils consists of a microalgal biomass production step that requires light, carbon dioxide, water, and inorganic nutrients. The latter are mainly nitrates, phosphates, iron, and some trace elements. Approximately half of the dry weight of microalgal biomass is carbon, which is typically derived from carbon dioxide. Therefore, producing 100 tons of algal biomass fixes roughly 183 tons of carbon dioxide. This carbon dioxide must be fed continually during daylight hours. It is often available at little or no cost (Chisti 2008). The optimal temperature for growing many microalgae is between 293 and 303 K. A temperature outside this range could kill or otherwise damage the cells.

There are three well-known methods to extract oil from algae: (1) expeller/press, (2) solvent extraction with hexane, and (3) supercritical fluid extraction. A simple process is to use a press to extract a large percentage (70 to 75%) of the oils from algae. Algal oil can be extracted using chemicals. The most popular chemical for solvent extraction is hexane, which is relatively inexpensive. Supercritical fluid extraction is far more efficient than traditional solvent separation methods. Supercritical fluids are selective, thus providing the high purity and product concentrations (Paul and Wise 1971). This method alone can allow one to extract almost 100% of the oils. In supercritical fluid CO_2 extraction, CO_2 is liquefied under pressure and heated to the point where it has the properties of both a liquid and a gas. This liquefied fluid then acts as the solvent in extracting the oil.

The lipid and fatty acid contents of microalgae vary in accordance with culture conditions. Algal oil contains saturated and monounsaturated fatty acids. The fatty acids exist in algal oil in the following proportions: 36% oleic (18:1), 15% palmitic (16:0), 11% stearic (18:0), 8.4% iso-17:0, and 7.4% linoleic (18:2). The high proportion of saturated and monounsaturated fatty acids in this alga is considered optimal from a fuel quality standpoint, in that fuel polymerization during combustion would be substantially less than what would occur with polyunsaturated fatty-acid-derived fuel (Sheehan *et al.* 1998). Table 6.1 shows the oil contents of some microalgae. Oil levels of 20 to 50% are quite common (Chisti 2007; Carlsson *et al.* 2007; Demirbas 2009a).

After oil extraction from algae, the remaining biomass fraction can be used as a high protein feed for livestock (Schneider 2006; Haag 2007). This gives further value to the process and reduces waste.

144 6 Biodiesel from Algae

Table 6.1 Oil contents of some microalgae

Microalga	Oil content (% dry wt)
Botryococcus braunii	25–75
Chlorella spp.	28–32
Crypthecodinium cohnii	20
Cylindrotheca spp.	16–37
Dunaliella primolecta	23
Isochrysis spp.	25–33
Monallanthus salina N	20
Nannochloris spp.	20–35
Nannochloropsis spp.	31–68
Neochloris oleoabundans	35–54
Nitzschia spp.	45–47
Phaeodactylum tricornutum	20–30
Schizochytrium spp.	50–77
Tetraselmis sueica	15–23

6.2.1 Production of Biodiesel from Algal Oils

Most current research on oil extraction is focused on microalgae to produce biodiesel from algal oil. The biodiesel from algal oil in itself is not significantly different from biodiesel produced from vegetable oils.

Dilution, microemulsification, pyrolysis, and transesterification are the four techniques applied to solve the problems encountered with high fuel viscosity. Of the four techniques, transesterification of oil into its corresponding fatty ester (biodiesel) is the most promising solution to the high viscosity problem. This is accomplished by mixing methanol with sodium hydroxide to make sodium methoxide. This liquid is then mixed into vegetable oil. The entire mixture then settles and glycerin is left on the bottom while methyl esters, or biodiesel, is left on top. Biodiesel can be washed with soap and glycerin using a centrifuge and then filtered. Kinematic viscosities of the fatty acid methyl esters vary from 3.23 to 5.61 mm^2/s (Knothe 2005). Methanol is preferred for transesterification because it is less expensive than ethanol (Graboski and McCormick 1998).

For production of biodiesel, a macroalga (*Cladophora fracta*) sample and a microalga (*Chlorella protothecoides*) sample were used in one study (Demirbas 2009b). Proximate analysis data and higher heating values of algae samples are given in Table 6.1. As seen in Table 6.2, the higher heating value of *Chlorella protothecoides* (25.1 MJ/kg) is also higher than that of *Cladophora fracta* (21.1 MJ/kg). Moisture content was determined by drying a 3- to 5-g sample at 378 K to constant weight (Demirbas 1999), ashing was carried out at 1,025 K for 2 h (Demirbas 2001), and protein content was determined by the block digestion method and ether-extractable intramuscular fat content by solvent extraction (Boccard *et al.* 1981). Table 6.3 shows the average chemical composition of algae samples. The oil proportion from the lipid fractions of *Chlorella protothecoides* is considerable higher than that of *Cladophora*

6.2 Biodiesel from Algal Oil

Table 6.2 Proximate analysis data and higher heating values of *Cladophora fracta* (an alga) and *Chlorella protothecoides* (a microalga), percent of dry-matter and ash-free basis (% daf)

Species of sample	Fixed carbon (% daf)	Volatile matter (% daf)	Higher Heating Value (MJ/kg)
Cladophora fracta	28.1	65.6	21.1
Chlorella protothecoides	39.6	54.6	25.1

Table 6.3 Average chemical composition of algae samples on a dry-matter basis (%)

Species of sample	Proteins	Carbohydrates	Lipids	Others
Cladophora fracta	52.3 ± 2.0	15.6 ± 0.9	14.2 ± 0.8	17.5 ± 0.9
Chlorella protothecoides	54.1 ± 2.2	13.7 ± 0.7	29.4 ± 1.5	4.8 ± 0.4

fracta (Demirbas 2009b). Figure 6.2 shows the production of biodiesel from algae.

Oils were obtained by extracting algae with hexane in a Soxhlet extractor for 18 h. Transesterification of algal oils was performed in a 100-mL cylinder using supercritical methanol according to earlier methods (Kusdiana and Saka 2001; Demirbas 2002). The fatty acids of the algal oils were fractionated into saturated, mo-

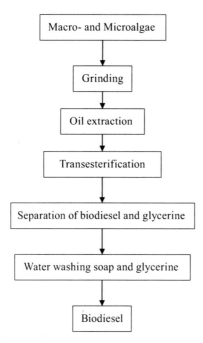

Figure 6.2 Production of biodiesel from algae

146　　6 Biodiesel from Algae

nounsaturated, polyunsaturated, and free forms by a preparative chromatographic thin layer on a glass plate coating with a 0.25-μm polyethanol succinate.

The fatty acid compositions of algal oils are given in Table 6.4. Fatty acids come in two varieties: saturated and unsaturated. Saturated fats come from animal products such as meat and dairy. Most vegetable oils are unsaturated. The properties of the various individual fatty esters that comprise biodiesel determine the overall fuel properties of the biodiesel fuel. As seen in Table 6.4, the average polyunsaturated fatty acids of *Chlorella protothecoides* (62.8%) are also higher than those of *Cladophora fracta* (50.9%). Algae generally produce a lot of polyunsaturates, which may present a stability problem since higher levels of polyunsaturated fatty acids tend to decrease the stability of biodiesel. However, polyunsaturates also have much lower melting points than monounsaturates or saturates; thus algal biodiesel should have much better cold weather properties than many other bio-oils (Demirbas 2009b).

Table 6.4 Fatty acid compositions of algal oils on a dry-matter basis (%)

Fatty acids in oils	*Cladophora fracta*	*Chlorella protothecoides*
Saturates	12.5 ± 0.7	10.8 ± 0.6
Monounsaturates	33.7 ± 1.6	24.1 ± 1.2
Polyunsaturates	50.9 ± 1.9	62.8 ± 2.5
Free	3.6 ± 0.3	2.6 ± 0.2

Xu *et al.* (2006) used *Chlorella protothecoides* (a microalga) for the production of biodiesel. Cells were harvested by centrifugation, washed with distilled water, and then freeze dried. The main chemical components of heterotrophic *C. protothecoides* were measured as in a previous study (Miao *et al.* 2004). Microalgal oil was prepared by pulverization of heterotrophic cell powder in a mortar and extraction with *n*-hexane.

Biodiesel was obtained from heterotrophic microalgal oil by acidic transesterification. Figure 6.3 shows the process flow schematic for biodiesel production (Xu *et al.* 2006). The optimum process combination was 100% catalyst quantity (based on oil weight) with 56:1 molar ratio of methanol to oil at a temperature of 303 K, which reduced product-specific gravity from an initial value of 0.912 to a final value of 0.864 in about 4 h of reaction time (Xu *et al.* 2006).

The technique of metabolic control through heterotrophic growth of *C. protothecoides* was applied, and the heterotrophic *C. protothecoides* contained a crude lipid content of 55.2%. To increase the biomass and reduce the cost of algae, corn powder hydrolysate instead of glucose was used as an organic carbon source in heterotrophic culture medium in fermenters. The result showed that cell density significantly increased under heterotrophic conditions, and the highest cell concentration reached 15.5 g/L. A large amount of microalgal oil was efficiently extracted from the heterotrophic cells using *n*-hexane and then transmuted into biodiesel by acidic transesterification (Xu *et al.* 2006).

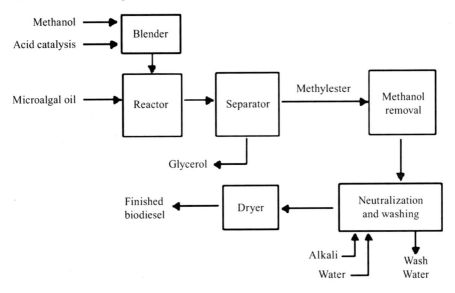

Figure 6.3 Process flow for biodiesel production

6.3 Potential of Microalgal Biodiesel

Algae biomass cultivation confers four important potential benefits that other sources don't have. First, algae biomass can be produced at extremely high volumes, and this biomass can yield a much higher percentage of oil than other sources. Second, algal oil has limited market competition. Third, algae can be cultivated on marginal land, fresh water, or sea water. Fourth, innovations to algae production allow it to become more productive while consuming resources that would otherwise be considered waste (Campbell 2008).

Biodiesel derived from oil crops is a potential renewable and carbon-neutral alternative to petroleum fuels. Unfortunately, biodiesel from oil crops, waste cooking oil, and animal fat cannot realistically satisfy even a small fraction of the existing demand for transport fuels. Microalgae appear to be the only source of renewable biodiesel that is capable of meeting the global demand for transport fuels.

Biodiesel production from biorenewable sources has a number of problems. First, most biorenewable sources, such as waste oil, animal fat, and vegetable oil, have a limited supply (Ma and Hanna 1999). Second, many of these sources have competitive uses, such as food or cosmetic production. Extensive use of renewable oils may cause other significant problems such as starvation in poor and developing countries (Demirbas 2007). Third, the resources that were used to create the biomass have competition with other uses, and this includes arable land. Third, because of the limited supply and competition, many sources of biomass have become increasingly expensive (Haag 2007).

Like plants, microalgae use sunlight to produce oils, but they do so more efficiently than crop plants. Oil productivity of many microalgae greatly exceeds the oil productivity of the best producing oil crops. Approaches to making microalgal biodiesel economically competitive with petrodiesel have been discussed (Chisti 2007). Biodiesel derived from green algae biomass has the potential for high-volume, cost-effective production (Campbell 2008).

Laboratory studies exploring methods to maximize both density and oil content have demonstrated that there is yet much unrealized potential. Xu *et al.* (2006) cultivated the algae *Chlorella prototothecoids* in a light-deprived, heterotrophic environment with inexpensive hydrolyzed corn starch as the sole food source. The algae were not only able to adapt to this environment, but they reached a high population density of 15.5 g/L.

Biodiesel from microalgae seems to be the only renewable biofuel that has the potential to completely displace petroleum-derived transport fuels without adversely affecting the food supply and other crop products. Most productive oil crops, such as oil palm, do not come close to microalgae in being able to sustainably provide the necessary amounts of biodiesel. Similarly, bioethanol from sugar cane is no match for microalgal biodiesel (Chisti 2008).

Microalgae contain lipids and fatty acids as membrane components, storage products, metabolites, and sources of energy. Algae present an exciting possibility as a feedstock for biodiesel, especially when you realize that oil was originally formed from algae.

In order to have an optimal yield, these algae need to have CO_2 in large quantities in the basins or bioreactors where they grow. Thus, the basins and bioreactors need to be coupled with traditional electricity-producing thermal power centers that produce CO_2 at an average rate of 13% of the total flue gas emissions. The CO_2 is put into the basins and assimilated by the algae. It is thus a technology that recycles CO_2 while also treating used water.

Algae can grow practically anywhere where there is enough sunshine. Some algae can grow in saline water. All algae contain proteins, carbohydrates, lipids, and nucleic acids in varying proportions. While the percentages vary with the type of algae, there are algae types that are comprised of up to 40% of their overall mass by fatty acids (Becker 1994). The most significant distinguishing characteristic of algal oil is its yield and, hence, its biodiesel yield. According to some estimates, the yield (per acre) of oil from algae is over 200 times the yield from the best-performing plant/vegetable oils (Sheehan *et al.* 1998). Microalgae are the fastest growing photosynthesizing organisms. They can complete an entire growing cycle every few days. Approximately 46 tons of oil/ha/year can be produced from diatom algae. Different algae species produce different amounts of oil. Some algae produce up to 50% oil by weight. The production of algae to harvest oil for biodiesel has not been undertaken on a commercial scale, but working feasibility studies have been conducted to arrive at the above number.

Microalgae are very efficient solar energy converters, and they can produce a great variety of metabolites (Chaumont 2005). The culture of algae can yield 30 to 50% oil (Chisti 2007; Dimitrov 2008). Oil supply is based on the theoretical claims that 47,000 to 308,000 L/ha/year of oil could be produced using algae. The calculated cost per barrel would be only $20 (Demirbas 2009a). Currently, a barrel of oil in the US market sells for over $100. Despite all the claims and research dating from the early 1970s to date, none of the projected algae and oil yields have been achieved (Patil *et al.* 2005). Algae, like all plants, require large quantities of nitrogen fertilizer and water, plus significant fossil energy inputs for the functioning system (Goldman and Ryther 1977). Harvesting the algae from tanks and separating the oil from the algae are difficult and energy-intensive processes (Pimentel *et al.* 2004; Pimentel 2008).

Fatty acids come in two varieties: saturated and unsaturated. Saturated fats come from animal products such as meat and dairy. Most vegetable oils are unsaturated. The properties of the various individual fatty esters that comprise biodiesel determine the overall fuel properties of the biodiesel fuel. Algae generally produce a lot of polyunsaturates, which may present a stability problem since higher levels of polyunsaturated fatty acids tend to decrease the stability of biodiesel. However, polyunsaturates also have much lower melting points than monounsaturates or saturates; thus algal biodiesel should have much better cold-weather properties than many other bio-oils (Demirbas 2009b). Algae are theoretically a very promising source of biodiesel. The lipid and fatty acid contents of microalgae vary in accordance with culture conditions. In some cases, lipid content can be enhanced by the imposition of nitrogen starvation or other stress factors. Which is the best species of algae for biodiesel? There is no one strain or species of algae that can be said to be the best in terms of oil yield for biodiesel. However, diatoms and secondly green algae have shown the most promise. *Scenedesmus dimorphus* is a unicellular alga in the class Chlorophyceae (green algae). While this is one of the preferred species for oil yield for biodiesel, one of the problems with *Scenedesmus* is that it is heavy and forms thick sediments if not kept in constant agitation. The strain known as *Dunaliella tertiolecta* has an oil yield of about 37% (organic basis). *D. tertiolecta* is a fast growing strain, which means it has a high CO_2 sequestration rate as well (Demirbas 2009a,b; Ozkurt 2009). Table 6.5 shows the yield of various plant oils.

Certain algae strains also produce polyunsaturated fatty acids (omega-3s) in the form of eicosapentaenoic acid (EPA) and docosahexaenoic acid (DHA) generally found in fish oils. Phototrophic microalgae are used to provide polyunsaturated fatty

Table 6.5 Yield of various plant oils

Crop	Oil (L/ha)
Algae	100,000
Castor	1,413
Coconut	2,689
Palm	5,950
Safflower	779
Soy	446
Sunflower	952

acids (omega-3 and omega-6) for aquaculture operations. These additional products greatly enhance the overall marketability and economics of producing algae (Volkman et al. 1989; Yaguchi et al. 1997; Vazhappilly and Chen 1998).

A selection of algae strains with the potential to be used for the production of oils for biofuel is presented in Table 6.6. A major current problem for the commercial viability of biodiesel production from microalgae is the low selling price of biodiesel (less than US$ 1.38/kg). Microalgal oils can potentially completely replace petroleum as a source of hydrocarbon feedstock for the petrochemical industry.

Table 6.6 Oil content in selected microalgal species

Species	Oil content (% dw)	Reference
Ankistrodesmus TR-87	28–40	Ben-Amotz and Tornabene 1985
Botryococcus braunii	29–75	Sheehan et al. 1998; Banerjee et al. 2002; Metzger and Largeau 2005
Chlorella spp.	29	Sheehan et al. 1998
Chlorella protothecoides (autotrophic/heterothrophic)	15–55	Xu et al. 2006
Cyclotella DI-35	42	Sheehan et al. 1998
Dunaliella tertiolecta	36–42	Kishimoto et al. 1994; Tsukahara and Sawayama 2005
Hantzschia DI-160	66	Sheehan et al. 1998
Isochrysis spp.	7–33	Sheehan et al. 1998; Valenzuela-Espinoza et al. 2002
Nannochloris	31 (6–63)	Ben-Amotz and Tornabene 1985; Negoro et al. 1991
Nannochloropsis	46 (31–68)	Hu et al. 2008
Nitzschia TR-114	28–50	Kyle and Gladue 1991
Phaeodactylum tricornutum	31	Sheehan et al. 1998
Scenedesmus TR-84	45	Sheehan et al. 1998
Stichococcus	33 (9–59)	Sheehan et al. 1998
Tetraselmis suecica	15–32	Sheehan et al. 1998; Zittelli et al. 2006; Chisti 2007
Thalassiosira pseudonana	(21–31)	Brown et al. 1996

6.4 Acceptability of Microalgal Biodiesel

The idea of producing biodiesel from microalgae that accumulate high amounts of oil was a main focus in the NREL project (Sheehan et al. 1998). Many species of algae accumulate large amounts of oils that to a large extent are made up of triacylglycerols consisting of three fatty acids bound to glycerol. The algal oil is converted into biodiesel through a transesterification process. Oil extracted from the algae is mixed with alcohol and an acid or a base to produce the fatty acid methylesters that makes up the biodiesel (Chisti 2007). A number of algae strains with good potential for making biodiesel were identified.

Some microalgae appear to be a suitable group of oleaginous microorganisms for lipid production (Chisti 2007). Microalgae have been suggested as potential candidates for fuel production because of a number of advantages including higher photosynthetic efficiency, higher biomass production, and higher growth rate compared to other energy crops (Milne *et al.* 1990; Dote *et al.* 1994; Minowa *et al.* 1995). Moreover, according to the biodiesel standard published by the American Society for Testing Materials (ASTM), biodiesel from microalgal oil is similar in properties to standard biodiesel and is also more stable according to their flash point values. Figure 6.4 shows a biodiesel product obtained from microalgae.

Figure 6.4 Biodiesel product obtained from microalgae

6.5 Economics of Biodiesel Production

There are small numbers of economic feasibility studies on microalgae oil (Richardson *et al.* 2009). Currently, microalgae biofuel has not been deemed economically feasible compared to the conventional agricultural biomass (Carlsson *et al.* 2007).

Critical and controversial issues are the potential biomass yield that can be obtained by cultivating macro- or microalgae and the costs of producing biomass and derived products. The basis of the estimates is usually a discussion of three parameters: photosynthetic efficiency, assumptions on scaleup, and long-term cultivation issues. For microalgae the productivity of raceway ponds and photobioreactors is limited by a range of interacting issues.

Typical productivity for microalgae in open ponds is 30 to 50 t/ha/y (Benemann and Oswald 1996; Sheehan *et al.* 1998). Several possible target areas to improve productivity in large-scale installations have been proposed (Benemann and Oswald 1996; Grobbelaar 2000; Suh and Lee 2003; Torzillo *et al.* 2003; Carvalho *et al.* 2006).

Harvesting costs contribute 20 to 30% to the total cost of algal cultivation, with the majority of the cost attributable to cultivation expenses. Genetic engineering, development of low-cost harvesting processes, improvements in photobioreactor, and integration of coproduction of higher-value products/processes are other alternatives in reducing algal oil production costs (Chisti 2007). The harvested algae then undergo anaerobic digestion, producing methane that could be used to produce electricity.

In commercial photobioreactors, higher productivities may be possible. Typical productivity for a microalga (*Chlorella vulgaris*) in photobioreactors is 13 to 150 (Pulz 2001). Photobioreactors require ten times more capital investment than open-pond systems. The estimated algal production cost for open-pond systems ($ 10/kg) and photobioreactors ($ 30 to $ 70/kg) is, respectively, two orders of magnitude higher and almost three orders of magnitude higher than conventional agricultural biomass (Carlsson *et al.* 2007). Assuming that biomass contains 30% oil by weight and carbon dioxide is available at no cost (flue gas), Chisti (2007) estimated the production cost for photobioreactors and raceway ponds at $ 1.40 and $ 1.81 per liter of oil, respectively. However, for microalgal biodiesel to be competitive with petrodiesel, algal oil should be less than $ 0.48/L (Chisti 2007).

It is useful to compare the potential of microalgal biodiesel with bioethanol from sugar cane, because on an equal energy basis, sugar cane bioethanol can be produced at a price comparable to that of gasoline (Bourne 2007). Bioethanol is well established for use as a transport fuel (Gray *et al.* 2006), and sugar cane is the most productive source of bioethanol (Bourne Jr. 2007). For example, in Brazil, the best bioethanol yield from sugar cane is 7.5 m^3/ha (Bourne Jr. 2007). However, bioethanol has only approx. 64% of the energy content of biodiesel. Therefore, if all the energy associated with 0.53 billion m^3 of biodiesel that the USA needs annually (Chisti 2007) were to be provided by bioethanol, nearly 828 million m^3 of bioethanol would be needed. This would require planting sugar cane over an area of 111 million ha, or 61% of total available US cropland.

Recovery of oil from microalgal biomass and conversion of oil into biodiesel are not affected by whether the biomass is produced in raceways or photobioreactors. Hence, the cost of producing the biomass is the only relevant factor for a comparative assessment of photobioreactors and raceways for producing microalgal biodiesel. If the annual biomass production capacity is increased to 10,000 t, the cost of production per kilogram reduces to roughly $ 0.47 and $ 0.60 for photobioreactors and raceways, respectively, because of economies of scale. Assuming that the biomass contains 30% oil by weight, the cost of biomass for providing a liter of oil would be something like $ 1.40 and $ 1.81 for photobioreactors and raceways, respectively (Chisti 2007).

Biodiesel from palm oil costs roughly $ 0.66/L, or 35% more than petrodiesel. This suggests that the process of converting palm oil into biodiesel adds about $ 0.14/L to the price of oil. For palm-oil-sourced biodiesel to be competitive with petrodiesel, the price of palm oil should not exceed $ 0.48/L, assuming no tax on biodiesel. Using the same analogy, a reasonable target price for microalgal oil is $ 0.48/L for algal diesel to be cost competitive with petrodiesel.

6.6 Improving Economics of Microalgal Biodiesel

Algae are among the fastest growing plants in the world, and about 50% of their weight is oil. That lipid oil can be used to make biodiesel for cars, trucks, and airplanes. Algae will some day be competitive as a source of biofuel.

Only renewable biodiesel can potentially completely displace liquid fuels derived from petroleum. The economics of producing microalgal biodiesel need to improve substantially to make it competitive with petrodiesel, but the level of improvement necessary appears to be attainable (Demirbas 2009b).

Biodiesel has great potential; however, the high cost and limited supply of renewable oils prevent it from becoming a serious competitor with petroleum fuels. As petroleum fuel costs rise and supplies dwindle, biodiesel will become more attractive to both investors and consumers. For biodiesel to become the alternative fuel of choice, it requires an enormous quantity of cheap biomass. Using new and innovative techniques for cultivation, algae may allow biodiesel production to achieve the price and scale of production needed to compete with, or even replace, petroleum (Campbell 2008).

It has been estimated that 0.53 billion m^3 of biodiesel would be needed to replace current US transportation consumption of all petroleum fuels (Chisti 2007). Neither waste oil nor seed oil can come close to meeting the requirement for that much fuel; therefore, if biodiesel is to become a true replacement for petroleum, a more productive source of oil such as algal oil is needed (Scott and Bryner 2006; Chisti 2007).

The cost of producing microalgal biodiesel can be reduced substantially by using a biorefinery-based production strategy, improving capabilities of microalgae through genetic engineering and advances in photobioreactor engineering. Like a petroleum refinery, a biorefinery uses every component of the biomass raw material to produce usable products (Chisti 2007).

6.7 Advantages and Disadvantages of Biodiesel from Algal Oil

Producing biodiesel from algae has been touted as the most efficient way to make biodiesel fuel. Algal oil processes into biodiesel as easily as oil derived from land-based crops. The difficulties in efficient biodiesel production from algae lie not in the extraction of the oil but in finding an algal strain with a high lipid content and fast growth rate that is not too difficult to harvest and a cost-effective cultivation system (i.e., type of photobioreactor) that is best suited to that strain.

Algae are the fastest growing plants in the world. Microalgae have much faster growth rates than terrestrial crops. The per-unit area yield of oil from algae is estimated to be between 18,927 and 75,708 L/acre/year; this is 7 to 31 times greater than the next best crop, palm oil, at 2,404 L/acre/year.

Algae are very important as a biomass source. Different species of algae may be better suited for different types of fuel. Algae can be grown almost anywhere,

Table 6.7 Advantages of biodiesel from algae oil

Rapid growth rates
Grows practically anywhere
A high per-acre yield (7 to 31 times greater than the next best crop, palm oil)
No need to use crops such as palms to produce oil
A certain species of algae can be harvested daily
Algae biofuel contains no sulfur
Algae biofuel is nontoxic
Algae biofuel is highly biodegradable
Algal oil extracts can be used as livestock feed and even processed into ethanol
High levels of polyunsaturates in algal biodiesel are suitable for cold weather climates
Can reduce carbon emissions based on where it's grown

Table 6.8 Disadvantages of biodiesel from algal oil

Produces unstable biodiesel with many polyunsturates
Biodiesel performs poorly compared to its mainstream alternative
Relatively new technology

even on sewage or salt water, and does not require fertile land or food crops, and processing requires less energy than the algae provides. Algae can be a replacement for oil-based fuels, one that is more effective. Algae consume CO_2 as they grow, so they could be used to capture CO_2 from power stations and other industrial plant that would otherwise go into the atmosphere. Tables 6.7 and 6.8 show the advantages and disadvantages of biodiesel from algal oil.

References

Aresta, M., Dibenedetto, A., Carone, M., Colonna, T., Fagale, C. 2005. Production of biodiesel from macroalgae by supercritical CO_2 extraction and thermochemical liquefaction. Environ Chem Lett 3:136–139.

Banerjee, A., Sharma, R., Chisti, Y., Banerjee, U.C. 2002. *Botryococcus braunii*: a renewable source of hydrocarbons and other chemicals. Crit Rev Biotechnol 22:245–279.

Becker, E.W. 1994. In: Baddiley, J. *et al.* (eds.). Microalgae: Biotechnology and Microbiology, Cambridge University Press, Cambridge, New York.

Ben-Amotz, A., Tornabene, T.G. 1985. Chemical profile of selected species of macroalgae with emphasis on lipids. J Phycol 21:72–81.

Benemann, J., Oswald, W.J. 1996. Systems and economic analysis of microalgae ponds for conversion of CO_2 to biomass. US DOE, Pittsburgh Energy Technology Center.

Boccard, R., Buchter, L., Casteels, E., Cosentino, E., Dransfield, E., Hood, D.E., Joseph, R.L., Macdougall, D.B., Rhodes, D.N., Schön, L., Tinbergen, B.J., Touraill, C. 1981. Procedures for measuring meat quality characteristic in beef production experiments. Report of a Working Group in the Commission of the European community's (CEC) Beef Production Research Programme, Livestock Prod Sci 8:385–397.

Bourne Jr., J.K. 2007. Biofuels: green dreams. Natl Geogr Mag October:41–59.

Bowman, M., Hilligoss, D., Rasmussen, S., Thomas, R. 2006. Biodiesel: a renewable and biodegradable fuel. Hydrocarbon Proc 2006:103–106.

References

Brown, M. R., Dunstan, G. A., Norwood, S. J., Miller, K. A. 1996. Effects of harvested stage and light on the biochemical composition of the diatom *Thalassiosira pseudonana*. J Phycol 32:64–73.

Campbell, M. N. 2008. Biodiesel: algae as a renewable source for liquid fuel. Guelph Eng J 1:2–7.

Cardone, M., Prati, M. V., Rocco, V., Senatore, A. 1998. Experimental analysis of performances and emissions of a diesel engines fuelled with biodiesel and diesel oil blends. Proceedings of MIS–MAC V, Rome, pp. 211–225 [in Italian].

Carlsson, A. S., van Bilen, J. B., Möller, R., Clayton, D. 2007. Mircro- and macro-algae: utility for industrial applications. http://www.epobio.net/pdfs/0709AquaticReport.pdf. Accessed June 2008.

Carvalho, A. P., Meireles, L. A., Malcata, F. X. 2006. Microalgal reactors: a review of enclosed system designs and performances. Biotechnol Prog 22:1490–1506.

Chaumont, D. 2005. Biotechnology of algal biomass production: a review of systems for outdoor mass culture. J Appl Phycol 5:593–604.

Chisti, Y. 2007. Biodiesel from Microalgae. Biotechnol Adv 25:294–306.

Chisti, Y. 2008. Biodiesel from microalgae beats bioethanol. Trends Biotechnol 26:126–131.

Crookes, R. J. 2006. Comparative bio-fuel performance in internal combustion engines. Biomass Bioenergy 30:461–468.

Demirbas, A. 1998. Fuel properties and calculation of higher heating values of vegetable oils. Fuel 77:1117–1120.

Demirbas, A. 1999. Proximate and heavy metal composition in chicken meat and tissues. Food Chem 67:27–31.

Demirbas, A. 2001. Mineral, protein, and fatty acids contents of hazelnut kernels. Energy Educ Sci Technol 7:37–43.

Demirbas, A. 2002. Biodiesel from vegetable oils via transesterification in supercritical methanol. Energy Convers Manage 43:2349–2356.

Demirbas, A. 2007. Importance of biodiese as transportation fuel. Energy Policy 35:4661–4670.

Demirbas, A. H. 2009a. Inexpensive oil and fats feedstocks for production of biodiesel. Energy Educ Sci Technol A 23:1–13.

Demirbas, A. 2009b. Production of biodiesel from algae oils. Energy Sources A 31:163–168.

Dimitrov, K. 2008. Green Fuel technologies: a case study for industrial photosythetic energy capture. Brisbane, Australia. Available from http://www.nanostring.net/Algae/.

Dote, Y., Sawayama, S., Inoue, S., Minowa, T., Yokoyama, Shin-ya. 1994 Recovery of liquid fuel from hydrocarbon-rich microalgae by thermochemical liquefaction. Fuel 73:1855–1857.

Goldman, J. C., Ryther, J. H. 1977. Mass production of algae: bio-engineering aspects. In:Mitsui, A. *et al.* (eds.). Biological Solar Energy Conversion, Academic, New York.

Graboski, M. S., McCormick, R. L. 1998. Combustion of fat and vegetable oil derived fuels in Diesel engines. Prog Energy Combust Sci 24:125–164.

Gray, K. A., Zhao, L., Emptage, M. 2006. Bioethanol. Curr Opin Chem Biol 10:141–146.

Grobbelaar, J. U. 2000. Physiological and technological considerations for optimising mass algal cultures. J Appl Phycol 12:201–206.

Haag, A. L. 2007. Algae bloom again. Nature 447:520–521.

Hu, Q., Sommerfeld, M., Jarvis, E., Ghirardi, M., Posewitz, M., Seibert, M., Darzins, A. 2008. Microalgal triacylglycerols as feedstocks for biofuel production: perspectives and advances. Plant J 54:621–639.

Kishimoto, M., Okakura, T., Nagashima, H., Minowa, T., Yokoyama, S., Yamaberi, K. 1994. CO_2 fixation and oil production using microalgae. J Ferment Bioeng 78:479–482.

Knothe, G. 2005. Dependence of biodiesel fuel properties on the structure of fatty acid alkyl esters. Fuel Proc Technol 86:1059–1070.

Kusdiana, D., Saka, S. 2001. Kinetics of transesterification in rapeseed oil to biodiesel fuels as treated in supercritical methanol. Fuel 80:693–698.

156 6 Biodiesel from Algae

Kyle, D. J., Gladue, R. M. 1991. Eicosapentaenoic acids and methods for their production. International Patent Application, Patent Cooperation Treaty Publication WO 91/14427, 3 October 1991.

Laforgia, D., Ardito, V. 1994. Biodiesel fuelled IDI engines: performances, emissions and heat release investigation. Biores Technol 51:53–59.

Ma, F., Hanna, M. A. 1999. Biodiesel production: a review. Biores Technol 70:1–15.

Metzger, P., Largeau, C. 2005. Botryococcus braunii: a rich source for hydrocarbons and related ether lipids. Appl Microbiol Biotechnol 66:486–496.

Milne, T. A., Evans, R. J., Nagle, N. 1990. Catalytic conversion of microalgae and vegetable oils to premium gasoline, with shape-selective zeolites. Biomass 21:219–232.

Minowa, T., Yokoya, S. Y., Kishimoto, M., Okakura, T. 1995. Oil production from algae cells of Dunaliella Tereiolata by direct thermochemical liquefaction. Fuel 74:1731–1738.

Miao, X. L.,Wu, Q. Y., Yang, C. Y. 2004. Fast pyrolysis of microalgae to produce renewable fuels. J Anal Appl Pyrol 71:855–863.

Negoro, M., Shioji, N., Miyamoto, K., Miura, Y. 1991. Growth of microalgae in high CO_2 gas and effects of sox and nox. Appl Biochem Biotechnol 28–29:877–886.

Ozkurt, I. 2009. Qualifying of safflower and algae for energy. Energy Educ Sci Technol A 23:145–151.

Patil, V., Reitan, K. I., Knudsen, G., Mortensen, L., Kallqvist, T., Olsen, E., Vogt, G., Gislerød, H. R. 2005. Microalgae as source of polyunsaturated fatty acids for aquaculture. Curr Topics Plant Biol 6:57–65.

Paul, P. F. M., Wise, W. S. 1971. The Principle of Gas Extraction. Mills and Boon, London.

Pimentel, D. (ed.). 2008. Biofuels, Solar and Wind as Renewable Energy Systems: Benefits and Risks. Springer, New York.

Pimentel, D., Berger, B., Filiberto, D., Newton, M., Wolfe, B., Karabinakis, B., Clark, S., Poon, E., Abbett, E., Nandagopal, S. 2004. Water resources: agricultural and environmental issues. Bioscience 54:909–918.

Pradeep, V., Sharma, R. K. 2007. Use of HOT EGR for nox control in a compression ignition engine fuelled with biodiesel from Jatropha oil. Renew Energy 32:1136–1154.

Pulz, O. 2001. Photobioreactors: production systems for phototrophic microorganisms. Appl Microbiol Biotechnol 57:287–293.

Rakopoulos, C. D., Antonopoulos, K. A., Rakopoulos, D. C., Hountalas, D. T., Giakoumis, E. G. 2006. Comparative performance and emissions study of a direct injection diesel engine using blends of diesel fuel with vegetable oils or bio-diesels of various origins. Energy Convers Manage 47:3272–3287.

Richardson, J. W., Outlaw, J. L., Allison, M. 2009. Economics of micro algae oil. 13th ICABR Conference on the Emerging Bio-Economy. Ravello, Italy, 17–20 June 2009.

Schneider, D. 2006. Grow your own?: Would the widespread adoption of biomass-derived transportation fuels really help the environment. Am Sci 94:408–409.

Scott, A., Bryner, M. 2006. Alternative fuels: rolling out next-generation technologies. Chem Week 20–27:17–21.

Sharif Hossain, A. B. M., Salleh, A., Boyce, A. N., Chowdhury, P., Naqiuddin, M. 2008. Biodiesel Fuel Production from Algae as Renewable Energy. Am J Biochem Biotechnol 4:250–254.

Sheehan, J., Dunahay, T., Benemann, J., Roessler, P. 1998. A Look Back at the U.S. Department of Energy's Aquatic Species Program – Biodiesel from Algae. National Renewable Energy Laboratory (NREL) Report: NREL/TP-580-24190. Golden, CO.

Suh, I. S., Lee, C. G. 2003. Photobioreactor engineering: design and performance. Biotechnol Bioproc Eng 8:313–321.

Torzillo, G., Pushparaj, B., Masojidek, J., Vonshak, A. 2003. Biological constraints in algal biotechnology. Biotechnol Bioproc Eng 8:338–348.

Tsukahara, K., Sawayama, S. 2005. Liquid fuel production using microalgae. J Jpn Petrol Inst 48:251–259.

References

Valenzuela-Espinoza, E., Millan-Nunez, R., Nunez-Cebrero, F. 2002. Protein, carbohydrate, lipid and chlorophyll alpha content in Isochrysis aff. galbana (clone T-Iso) cultured with a low cost alternative to the f/2 medium. Aquacult Eng 25:207–216.

Vazhappilly, R., Chen, F. 1998. Heterotrophic production potential of omega-3 polyunsaturated fatty acids by microalgae and algae-like microorganisms. Botan Marina 41:553–558.

Volkman, J. K., Jeffrey, S. W., Nichols, P. D., Rogers, G. I., Garland, C. D. 1989. Fatty acid and lipid composition of 10 species of microalgae used in mariculture. J Exp Marine Biol Ecol 128:219–240.

Xu, H., Miao, X., Wu, Q. 2006. High quality biodiesel production from a microalga Chlorella protothecoides by heterotrophic growth in fermenters. J Biotechnol 126:499–507.

Yaguchi, T., Tanaka, S., Yokochi, T., Nakahara, T., Higashihara, T. 1997. Production of high yields of docosahexaenoic acid by *Schizochytrium sp.* strain SR21. J Am Oil Chem Soc 74:1431–1434.

Zhang, X. 1996. Biodegradability of biodiesel in the aquatic and soil environments. Ph.D. dissertation, Dept. of Biol. and Agr. Engr., University of Idaho, Moscow, ID.

Zhang, X., Peterson, C., Reece, D., Haws, R., Moller, G. 1998. Biodegradability of biodiesel in the aquatic environment. Trans ASAE 41:1423–1430.

Zittelli, G. C., Rodolfi, L., Biondi, N., Tredici, M. R. 2006. Productivity and photosynthetic efficiency of outdoor cultures of Tetraselmis suecica in annular columns. Aquaculture 261:932–943.

Chapter 7
Biorefineries

7.1 Introduction

Energy needs are increasing continuously because of increases in industrialization and population. The growth of the world's energy demand raises urgent problems. The larger part of petroleum and natural gas reserves is located within a small group of countries. Today's energy system is unsustainable because of equity issues as well as environmental, economic, and geopolitical concerns that have implications far into the future. Bioenergy is one of the most important components of greenhouse-gas-emissions mitigation and fossil–fuel replacement (Goldemberg 2000; Dincer 2008). Renewable energy is one of the most efficient ways to achieve sustainable development.

Plants use photosynthesis to convert solar energy into chemical energy, which is stored in the form of oils, carbohydrates, proteins, etc. This plant energy is converted into biofuels. Hence biofuels are primarily a form of solar energy. For biofuels to succeed at replacing large quantities of petroleum fuel, the feedstock availability needs to be as high as possible. There is an urgent need to design integrated biorefineries that are capable of producing transportation fuels and chemicals.

In recent years, the recovery of liquid transportation biofuels from biorenewable feedstocks has became a promising method. The biggest difference between biorenewable and petroleum feedstocks is oxygen content. Biorenewables have oxygen levels from 10 to 44%, while petroleum has essentially none, making the chemical properties of biorenewables very different from those of petroleum (Demirbas 2008; Balat 2009). For example, biorenewable products are often more polar; some easily entrain water and can therefore be acidic.

There are two global transportation fuels – gasoline and diesel fuel. The main transportation fuels that can be obtained from biomass using different processes are sugar ethanol, cellulosic ethanol, grain ethanol, biodiesel, pyrolysis liquids, green diesel, green gasoline, butanol, methanol, syngas liquids, biohydrogen, algae diesel, algae jet fuel, and hydrocarbons. Renewable liquid biofuels for transportation have recently attracted considerable attention in various countries around the world be-

A. Demirbas, M. Fatih Demirbas, *Algae Energy*
DOI 10.1007/978-1-84996-050-2, © Springer 2010

cause of their renewability, sustainability, widespread availability, and biodegradability, as well as for their potential role in regional development, rural manufacturing jobs, and the reduction of greenhouse gas emissions (Demirbas 2008). Transportation fuels, both petroleum-based and biorenewable, are given in Figure 7.1.

The term biofuel or biorenewable fuel (refuel) is referred to as solid, liquid, or gaseous fuels that are predominantly produced from biomass. Liquid biofuels being considered the world over fall into the following categories: (a) bioalcohols, (b) vegetable oils and biodiesels, and (c) biocrude and synthetic oils. Biofuels are important because they replace petroleum fuels. It is expected that the demand for biofuels will rise in the future. Biofuels are substitute fuel sources for petroleum; however, some still include a small amount of petroleum in the mixture. It is generally considered that biofuels address many concerns, including sustainability, reduction of greenhouse gas emissions, regional development, social structure and agriculture, and supply security. Biofuels, among other sources of renewable energy, are attracting interest as alternative to fossil diesel. With an increasing number of governments now supporting this cause in the form of mandates and other policy initiatives, the biofuel industry is poised to grow at a phenomenal rate (Balat 2007; Demirbas 2002, 2003, 2007; Demirbas and Karslioglu 2007; Khoiyangbam 2008; Chhetri and Islam 2008).

Policy drivers for biorenewable liquid biofuels have attracted support for rural development and economic opportunities for developing countries (Keskin 2009). The EU ranks third in biofuel production worldwide, behind Brazil and the USA. In Europe, Germany is the largest and France the second largest producer of biofuels.

The term modern biomass is generally used to describe traditional biomass use through efficient and clean combustion technologies and sustained supply of

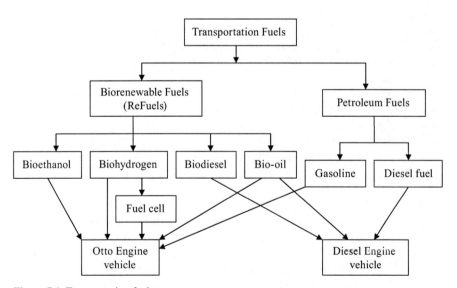

Figure 7.1 Transportation fuels

7.1 Introduction

biomass resources, environmentally sound and competitive fuels, heat, and electricity using modern conversion technologies. Biomass, as an energy source, has two striking characteristics. First, biomass is the only renewable organic resource to exist in abundance. Second, biomass fixes carbon dioxide in the atmosphere by photosynthesis. Direct combustion and cofiring with coal for electricity production from biomass holds great promise. Biomass thermochemical conversion technologies such as pyrolysis and gasification are certainly not the most important options at present; combustion is responsible for over 97% of the world's bioenergy production. Ethanol and fatty acid (m)ethylester (biodiesel), as well as diesel produced from biomass by Fischer–Tropsch synthesis (FTS), are modern biomass-based transportation fuels. Liquid transportation fuels can be economically produced by biomass-integrated gasification Fischer–Tropsch (BIG-FT) processes. Modern biomass produced in a sustainable way excludes traditional uses of biomass as fuel wood and includes electricity generation and heat production, as well as transportation fuels, from agricultural and forest residues and solid waste. On the other hand, traditional biomass is produced in an unsustainable way and is used as a noncommercial source – usually with very low efficiencies for cooking in many countries. Biomass energy potentials and current use in different regions are given in Table 7.1 (Parikka 2004).

Like a petroleum refinery, a biorefinery uses every component of the biomass raw material to produce usable products. Bio-based products are prepared for economic use by an optimal combination of different methods and processes (physical, chemical, biological, and thermal). Therefore, basic biorefinery concepts must be developed. A biorefinery, as a new approach, is a processing plant where biomass feedstocks are converted and extracted into a spectrum of valuable products. Biorefining refers to fractionating biomass into various separated products that possibly undergo further chemical, biochemical, biological, and thermochemical processing and separation. The molecules produced by biorefining can be obtained using thermal, chemical, mechanical, enzymatic, or microbial processes, and they can be used in transport fuels, therapeutics, food additives, or as secondary chemicals with a range of applications. By coproducing chemicals, the production costs of secondary energy carriers (fuels, heat, power) could potentially become more profitable, especially

Table 7.1 Biomass energy potentials and current use in different regions

Biomass potential	North America	Latin America	Asia	Africa	Europe	Middle East	Former USSR	World
Woody biomass	12.8	5.9	7.7	5.4	4.0	0.4	5.4	41.6
Energy crops	4.1	12.1	1.1	13.9	2.6	0.0	3.6	37.4
Straw	2.2	1.7	9.9	0.9	1.6	0.2	0.7	17.2
Other	0.8	1.8	2.9	1.2	0.7	0.1	0.3	7.6
Potential sum (EJ/a)	19.9	21.5	21.4	21.4	8.9	0.7	10.0	103.8
Use (EJ/a)	3.1	2.6	23.2	8.3	2.0	0.0	0.5	39.7
Use/potential (%)	16	12	108	39	22	7	5	38

when biorefining is integrated into the existing chemical, material, and power industries. Coproduction of bioproducts, materials, chemicals, transportation fuels, power, or heat in technically, economically, and ecologically fully optimized integrated biorefinery systems will be required.

7.2 Definitions of Biorefinery

A biorefinery is a facility that integrates biomass conversion processes and equipment to produce fuels, power, and value-added chemicals from biomass. The biorefinery concept is analogous to today's crude oil refinery, which produces multiple fuels and products from petroleum. Biorefinery refers to the conversion of biomass feedstock into a host of valuable chemicals and energy with minimal waste and emissions. In a broad definition, biorefineries process all kinds of biomass (all organic residues, energy crops, and aquatic biomass) into numerous products (fuels, chemicals, power and heat, materials, and food and feed). Figure 7.2 shows a schematic diagram of a biorefinery.

A biorefinery is a conceptual model for future biofuel production where both fuels and high-value coproduct materials are produced. Biorefineries can simultaneously produce biofuels as well as bio-based chemicals, heat, and power. Biorefiner-

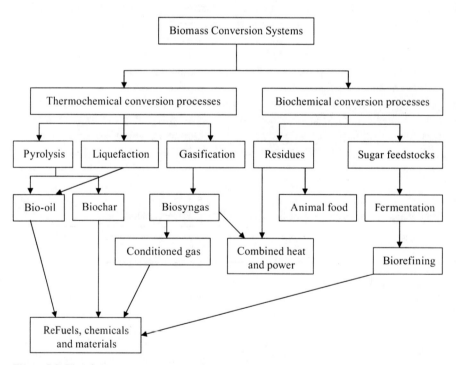

Figure 7.2 Biorefinery concept

7.2 Definitions of Biorefinery

ies present more cost-effective options where bio-based chemicals are coproducts of liquid fuel. Future biorefineries would be able to mimic the energy efficiency of modern oil refining through extensive heat integration and coproduct development. Heat that is released from some processes within the biorefinery could be used to meet the heat requirements for other processes in the system (WI 2007).

Reliable designs of equipment for the thermochemical stages of biomass conversion have yet to be widely demonstrated and shown capable of continuously producing synthesis gas of the required quality. Synthesis gas has to be free of nitrogen; this requires the use of oxygen (expensive), a pyrolytic process optimized for gas, or a multistage process. The main nontechnical barriers to acceptance of liquid biofuels, especially in the transport sector, relate to the costs of production, available markets, taxation policies, and legislation, as well as blending and distribution. From an economic point of view the use of agricultural crops, without subsidy, is too expensive to produce either bioethanol or biodiesel at a price competitive with untaxed petrol or diesel fuel, while processes for using lower cost lignocellulosic materials have not been perfected. There are a number of pilot-scale and demonstration plants operating, under development, or planned with fully commercial developments not expected for another decade or two (GBRP 2007).

Developed and developing countries clearly have different goals in the development of biomass energy. Biomass energy is promoted in developed countries as a replacement for fossil fuels, particularly in the transportation sector, whereas its use serves basic livelihood purposes in developing countries. Combining higher-value products with higher-volume energy production and employing any combination of conversion technologies has the greatest potential for making fuels, chemicals, materials, and power from biomass competitive. Obtaining modern biofuels, biopower, and bioproducts from biomass can be realized only in integrated biorefineries. This chapter reviews current biorenewable fuel valorization facilities as well as the future importance of biorefineries. The development of biorefinery technologies is important, and these technologies are also very promising.

Broadly speaking, the term biorefinery can be thought of as a concept of multiple products from various biomass feedstocks. A biorefinery processes biomass into value-added product streams. In theory, anything that uses biomass and makes more than one product is a biorefinery. A biorefinery is analogous to a petroleum refinery processing a range of crude oils. This very simple definition captures a wide range of existing, emerging, and advanced process concepts. Examples of existing biorefineries include corn processors and pulp and paper mills.

The concept is analogous to a combined use of fluid catalytic cracking, thermal cracking, and hydrocracking technology to convert the higher-boiling-range fractions of crude oil into more useful lower-boiling-range products. Just as few petroleum refineries use all available conversion technologies, biorefineries too will use only those technology platforms that are most cost effective for converting a certain type of biomass into a certain collection of desired end products. For crops and agricultural waste, it would be better to convert the biomass into bio-oil near the farm and transport the high-density bio-oil to a central facility for processing rather than transporting the low-density biomass.

The biorefinery concept attempts to apply to biomass conversion the methods that have been applied to the refining of petroleum. The goal is to maximize the value of the products obtained from the biomass. The goal of the integrated biorefinery program area is to support the establishment of integrated biorefineries through partnerships with industry and academia.

7.2.1 Main Technical and Nontechnical Gaps and Barriers to Biorefineries

There are a lot of technical and nontechnical gaps and barriers related to the implementation and commercialization of the biorefinery. Current technical barriers with the use of energy crops are associated with the cost of production and difficulties in harvesting and storing the material grown, especially for annual or other crops that have to be harvested within a narrow time period in the autumn. Transportation costs are of prime importance when calculating the overall cost of biomass; hence local or regional production of biomass is most favorable. Other technical problems associated with growing energy crops include provision of nutrients and control of pests and disease.

The major nontechnical barriers are restrictions or prior claims on use of land (food, energy, amenity use, housing, commerce, industry, leisure, or designated areas of natural beauty, special scientific interest, etc.), as well as the environmental and ecological effects of large areas of monoculture. For example, vegetable oils are a renewable and potentially inexhaustible source of energy with an energy content close to that of diesel fuel. On the other hand, extensive use of vegetable oils may cause other significant problems such as starvation in developing countries. Vegetable oil fuels are not acceptable because they were more expensive than petroleum fuels.

In addition to the technical challenges of commercializing advanced biorefineries, there are also large infrastructure barriers. These barriers are associated with the development of new agricultural infrastructure for the collection and storage of crop wastes. An integrated feedstock supply system must be developed that can supply the feedstock needs in a sustainable fashion at a reasonable cost. Infrastructure issues could be as significant as the technical issues when considering overall production costs.

7.3 Historical Perspective

The process of converting biorenewables into useful and higher-value compounds is not new. For example, sugar cane has been used in the production of bioethanol since 6000 BC (Demirbas 2009). Lactic acid was first discovered by C. W. Scheele in 1780. Its production was developed into an industrial-level fermentation process

7.3 Historical Perspective

by A. Boehringer in Germany in 1895, with subsequent uses in the food, leather, and textile industries. Even to this day, lactic acid is used as a preservative, flavor enhancer, and aciduant in the food industry.

Ethanol has been used by humans since prehistory as the intoxicating ingredient of alcoholic beverages. The ancient Egyptians produced alcohol by naturally fermenting vegetative materials. Also in ancient times, the Chinese discovered the art of distillation, which increases the concentration of alcohol in fermented solutions. Dried residues on 9000-year-old pottery have been found in China. Its isolation as a relatively pure compound was first achieved by Persian alchemist Zakariya Razi. Ethanol was first prepared synthetically in 1826 through the independent efforts of Henry Hennel in Great Britain. Ethanol was used as lamp fuel in the USA as early as 1840. In the 1970s, however, the ethanol industry began to reemerge when ethanol was used as a fuel extender during gasoline shortages caused by the OPEC oil embargoes.

Ethanol has been used in Germany and France as early as 1894 by the then-incipient industry of internal combustion engines. Brazil has utilized ethanol as a fuel since 1925. By that time, the production of ethanol was 70 times bigger than the production and consumption of petroleum.

The emergence of transesterification can be dated back to as early as 1846 when Rochieder described glycerol preparation through the ethanolysis of castor oil. Since that time alcoholysis has been studied in many parts of the world. Transesterification of a vegetable oil was conducted as early as 1853, by scientists E. Duffy and J. Patrick, many years before the first diesel engine became functional. Life for the diesel engine began in 1893 when the famous German inventor Dr. Rudolph Diesel published a paper entitled "The theory and construction of a rational heat engine." What the paper described was a revolutionary engine in which air would be compressed by a piston to a very high pressure, thereby causing a high temperature. Dr. R. Diesel designed the original diesel engine to run on vegetable oil.

Dr. Diesel's primary model, a single 3-m iron cylinder with a flywheel at its base, ran on its own power for the first time in Augsburg, Germany on 10 August 1893. (In remembrance of this event, 10 August has been declared International Biodiesel Day.) This engine stood as an example of Diesel's vision because it was powered by peanut oil, a biofuel, though not strictly biodiesel, since it was not transesterified. He believed that the utilization of a biomass fuel was the real future of his engine. In a 1912 speech, Dr. Diesel said, "The use of vegetable oils for engine fuels may seem insignificant today, but such oils may become, in the course of time, as important as petroleum and the coal-tar products of the present time."

The use of vegetable oils as an alternative renewable fuel competing with petroleum was proposed in the early 1980s. The advantages of vegetable oils as diesel fuel are its portability, widespread availability, renewability, higher heat content (about 88% of no. 2 petroleum diesel fuel), lower sulfur content, lower aromatic content, and biodegradability. The energy supply concerns of the 1970s renewed interest in biodiesel, but commercial production did not begin until the late 1990s.

Since the 1980s, biodiesel plants have opened in many European countries, and some cities have run buses on biodiesel, or blends of petro and bio diesels. More recently, Renault and Peugeot have approved the use of biodiesel in some of their truck engines. Recent environmental and domestic economic concerns have prompted a resurgence in the use of biodiesel throughout the world. In 1991, the European Community (EC) proposed a 90% tax deduction for the use of biofuels, including biodiesel. Biodiesel plants are now being built by several companies in Europe; each of these plants will produce up to 5.67 million liters of fuel per year. The EU accounted for nearly 89% of all biodiesel production worldwide in 2005.

A historical outline of the technology and industrial resources related to biodiesel must include discussions on the beginnings of sugar production, starch hydrolysis, wood saccharification, formation and separation of furfural, cellulose and pulp production, synthesis of levulinic acid, lipids, vanillin from lignin, and lactic acid fermentation. During the industrial revolution, coal became the energy source of preference, displacing biomass fuels such as wood. Since then a steady migration toward fossil fuels has continued, moving further away from biomass, not only for energy but also for sources of chemicals used to make everyday items. An excellent example of this is furfural, which can be obtained from oat hulls. Some current aspects of biorefinery research and development since the early 1990s have been presented, revealing that integrated processes, biomass refinery technology, and biorefinery technology have become objects of research and development. Table 7.2 shows the main biorefinery demonstration projects: plant capacity, planned or under construction.

Many of the currently used bio-based industry products are the results of direct physical or chemical treatment and processing of biomass: cellulose, starch, oil, protein, lignin, and terpenes. Also, with the help of biotechnological processes and methods, feedstock chemicals, such as ethanol, buthanol, acetone, lactic and itaconic acids, and amino acids, are produced.

The topic of biorefineries as a means of processing industrial material and efficiently utilizing renewable products is well known and applied worldwide, in almost every developed and emerging country. This trend is motivated by the rising cost of oil (petroleum) and the need to move away from petrochemical-based systems. The sustainability of integrated biorefining systems includes aspects of agriculture/forestry ecosystem modeling and analysis, as well as technical and economic considerations underlining the importance of biorefineries and the growing demand for new products.

Furthermore, the technology of sugar refining, starch production, and oil milling, the separation of proteins as feed, and the extraction of chlorophyll for industrial use with alfalfa as raw material were of great historical importance. But processes like wet grinding of crops and biotechnological processes such as the production of ethanol as well as acetic, lactic, and citric acids also used to be of fundamental importance in the 19th and 20th centuries. Figure 7.3 shows global trends in biofuel production by region.

Today's biorefinery technologies are based on (a) utilization of the whole plant or complex biomass and (b) the integration of traditional and modern processes for utilization of biological raw materials (Kamm *et al.* 2005). In the 19th and early 20th centuries, large-scale utilization of renewable resources was focused on pulp and paper production from wood, saccharification of wood, nitration of cellulose

7.3 Historical Perspective

Table 7.2 Biorefinery demonstration projects: plant capacity, planned or under construction

Name	Location	Participants	Annual capacity, million liters
Iogen	Shelley ID, USA	Iogen Corporation, Goldman Sachs, Royal Dutch Shell Oil Company; others	68 (in first plant) 945 (in future plants) cellulosic ethanol
BlueFire Ethanol	Lancaster, CA, USA	BlueFire Ethanol Fuels, Inc.	121 Cellulosic ethanol
Poet	Sioux Falls, SD, USA	Broin Companies	246 Corn ethanol
Ecofin, LLC	Washington County, KY, USA	University of Cincinnati, University of Kentucky, Alltech, Inc.	5 Corncob ethanol
ICM, Incorporated	St. Joseph, MO, USA	AGCO Eng; NCAUR-ARS-Peoria; CERES, Inc.; Edens. Sys. Corp.; Novozymes NA, Inc.; Sun EtOH, Inc.; others	5.7 Lignocellulosic ethanol
Lignol Innovations	Canada's Lignol Energy Corporation	Commerce City, CO, USA	9.55 Cellulosic ethanol
Royal Nedalco	Rotterdam Area, Netherlands	Ethanol Reach Association	227 Cellulosic ethanol
Mascoma	Rome, NY, USA	Mascoma Corporation, DuPont-Danisco	1.9 Cellulosic ethanol
Pacific Ethanol	Boardman, OR, USA	Pacific Ethanol, Inc.	10.2 Lignocellulosic ethanol
Abengoa Bioenergy	Colwich, KS, USA	Abengoa Bioenergy LLL	95 Lignocellulosic ethanol
Verenium	Jennings, LA, USA	Verenium Corporation	5.7 Cellulosic ethanol
RSE Pulp	Old Town, ME, USA	RSE Pulp & Chemical Llc Company	8.7 Cellulosic ethanol
Flambeau	Park Falls, WI, USA	Flambeau LLC	53.7 Lignocellulosic ethanol
NewPage	Wisconsin Rapids, WI, USA	NewPage Corporation	48.4 Biofuels, FT liquids
Choren	Freiberg, Germany	Choren Industries	5.7 Biofuels, FT liquids
Alico	LaBelle, FL, USA	Alico, Inc.; Bioeng. Res., Inc. Fayetteville, AR; Washington Group Intl. of Boise, ID; Emmaus Found., Inc.; others	28,4 Ethanol, electricity, ammonia, hydrogen
Range Fuels	Broomfield, CO, USA	Range Fuels, Inc.	4.5 Biofuels, FT liquids
Flambeau River	Jennings, LA, USA	Flambeau River BioFuels	24.6 Biofuels, FT liquids

Figure 7.3 Global trends in biofuel production by region

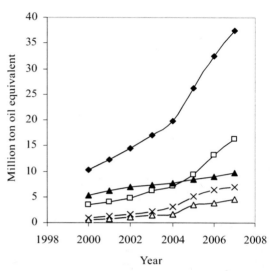

for guncotton and viscose silk, production of soluble cellulose for fibers, fat curing, and the production of furfural for Nylon. Advanced biorefineries are envisioned as the foundation of a new bioindustry. By exploiting new chemical, biological, and mechanical technologies, they offer the promise of greatly expanding the use of renewable plant-based materials, as well as a means of transitioning to a more energy-efficient and environmentally sustainable chemical and energy economy.

7.4 Petroleum Refinery and Biorefinery

In 1858, Dr. Abraham Gesner, a Canadian physician and amateur geologist, developed and patented the extraction of a lamp fuel from asphalt rock, which he named kerosene (Nova Scotia Museum 2008). At that time, kerosene, which was an extremely high-value lighting fuel for lamps, was the primary product of petroleum refining. For a while, distillation of kerosene for lamps was the mainstay of the new petroleum industry. Gasoline was merely a byproduct of kerosene production from crude oil, and until the early 1900s there was no significant demand for it. The first petrochemical, aside from carbon black manufactured on an industrial scale, was isopropyl alcohol, made by Standard Oil of New Jersey in 1920.

When a mixture of two liquids of different boiling points is heated to its boiling point, the vapor contains a higher mole fraction of the liquid with the lower boiling point than the original liquid, i.e., the vapor is enriched in the more volatile component. If this vapor is now condensed, the resultant liquid has also been enriched

7.4 Petroleum Refinery and Biorefinery

in the more volatile component. This is the principle of batch fractional distillation, and in a distillation column many, many such cycles are performed continuously, allowing almost complete separation of liquid components. A generalized distillation column is shown in Figure 7.4. The first step in the refining of crude oil, whether in a simple or a complex refinery, is the separation of the crude oil into fractions (fractionation or fractional distillation). These fractions are mixtures containing hydrocarbon compounds whose boiling points lie within a specified range.

Crude oil is a complex mixture that is between 50 and 95% hydrocarbon by weight. The first step in refining crude oil involves separating the oil into different hydrocarbon fractions by distillation. An oil refinery cleans and separates the crude oil into various fuels and byproducts, including gasoline, diesel fuel, heating oil, and jet fuel. The main crude oil components are listed in Table 7.3. Since various components boil at different temperatures, refineries use a heating process called distillation to separate the components. For example, gasoline has a lower boiling point than kerosene, allowing the two to be separated by heating to different temperatures. Another important job of the refineries is to remove contaminants from the oil, for example, sulfur from gasoline or diesel to reduce air pollution from the automobile exhausts. After processing at the refinery, gasoline and other liquid products are usually shipped out through pipelines, which are the safest and cheapest way to move large quantities of petroleum across land (Demirbas 2009).

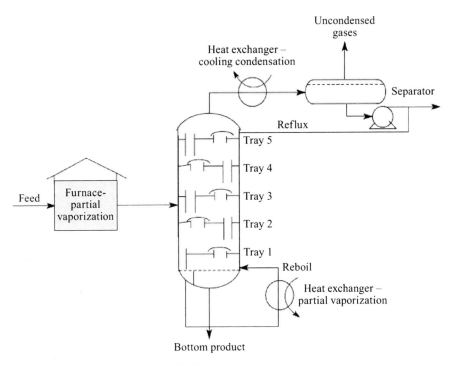

Figure 7.4 A generalized fractional distillation column

Table 7.3 Main crude oil fractions

Component	Boiling range, K	Number of carbon atoms
Natural gas	< 273	C_1 to C_4
Liquefied petroleum gas	231–273	C_3 to C_4
Petroleum ether	293–333	C_5 to C_6
Ligroin (light naphtha)	333–373	C_6 to C_7
Gasoline	313–478	C_5 to C_{12}, and cycloalkanes
Jet fuel	378–538	C_8 to C_{14}, and aromatics
Kerosene	423–588	C_{10} to C_{16}, and aromatics
No. 2 diesel fuel	448–638	C_{10} to C_{22}, and aromatics
Fuel oils	> 548	C_{12} to C_{70}, and aromatics
Lubricating oils	> 673	$> C_{20}$
Asphalt or petroleum coke	Nonvolatile residue	Polycyclic structures

The refining of heavy oil requires extracting and thorough chemical, engineering, and computing processes. Before the actual refining begins, the stored heavy crude oil is cleaned of contaminants such as sand and water.

Industrial distillation is typically performed in large, vertical, steel cylindrical columns known as distillation towers or distillation columns with diameters ranging from about 65 cm to 11 m and heights ranging from about 6 to 60 m or more. To improve the separation, the tower is normally provided inside with horizontal plates or trays, or the column is packed with a packing material. To provide the heat required for the vaporization involved in distillation and also to compensate for heat loss, heat is most often added to the bottom of the column by a reboiler. Large-scale industrial fractionation towers use reflux to achieve more efficient separation of products. Reflux refers to the portion of the condensed overhead liquid product from a distillation tower that is returned to the upper part of the tower. Inside the tower, the downflowing reflux liquid provides cooling and partial condensation of the upflowing vapors, thereby increasing the efficacy of the distillation tower. There are generally 25 to 45 plates or trays in a distillation tower. Each of the plates or trays is at a different temperature and pressure. The stage at the tower bottom has the highest pressure and temperature. Progressing upwards in the tower, the pressure and temperature decrease for each succeeding stage. Another way of improving the separation in a distillation column is to use a packing material instead of trays.

Three major refinery processes change crude oil into finished products: (1) separation, (2) conversion, and (3) purification. The first step is to separate the crude oil into its naturally occurring components. This is known as separation and is accomplished by applying heat through a process called distillation. Separation is performed in a series of distillation towers. The conversion processes have focused on reducing the length of some hydrocarbon chains. The primary purpose of conversion is to convert low valued heavy oil into high valued petrol. For example, catalytic reforming is a conversion process. The purpose of the reformer is to increase the octane number of the petrol blend components. Once crude oil has been through separation and conversion, the resulting products are ready for purification,

7.4 Petroleum Refinery and Biorefinery

which is principally sulfur removal. Common process units found in an oil refinery are presented in Table 7.4.

Petroleum refining is somewhat analogous to biorefining. Although biorefineries utilize different processing technologies, they separate and isolate components of biomass for the production of energy fuels, chemicals, and materials. Biorefineries can be designed and built to produce desired outputs from the processing of a wide variety of biorenewable materials. These biorefineries will adopt and integrate a range of materials handling and preprocessing equipment, thermochemical and biochemical conversion technologies, and new extraction and purification sciences to produce a range of intermediate products, while using less energy and reducing effluents and emissions. The scale of the biorefining operations will range from medium-sized to very large (equivalent in size to existing chemical plants and pulp and paper mills). Adoption of biorefineries and related processes and product technologies depends on available research, development, and prevailing regulations during design and construction.

In some ways, biorefineries are analogous to oil refineries. Oil refineries take crude oil and fractionate it into many different useful parts. This is done using a simple chemical distillation process. Biomass, like oil, consists of many different fractions that are separated and made into useful products in biorefineries. However, the processes involved in fractionating biomass are more complex than those used in oil refineries. Another important difference between biorefineries and oil refineries is their size. The term biorefinery was coined to describe future processing complexes that will use renewable agricultural residues, plant-based starch, and lignocellulosic materials as feedstocks to produce a wide range of chemicals, fuels, and bio-based materials. Biorefineries will most likely be limited in size, because biomass must be produced and transported economically from a limited catchment area. In contrast, oil is drilled and transported all over the world for processing.

Biomass can be processed into plastics, chemicals, fuels, heat, and power in a biorefinery. High-value components, for example essential oils, drugs, or fibers, can be recovered as a preprocessing step, with the remaining materials then processed downstream. Processing technologies are most advanced for chemicals and fuels. Biorefineries vary from small single-process plants to large multiprocess sites. Larger biorefineries will be able to integrate different technologies to obtain maximum value from biomass feedstocks.

A biorefinery is an integrated plant producing multiple value-added products from a range of renewable feedstocks. This innovative approach responds to changing markets for traditional forest products as well as new products such as energy, chemicals, and materials. The range of feedstocks, processes, and potential products is large; each combination of feedstock, process, and product is characterized by its own unique combination of technical and economic opportunities, emerging technologies, and barriers.

Table 7.5 shows the classification of biorefineries based on their feedstocks. A forest biorefinery will use multiple feedstocks including harvesting residues, extracts from effluents, and fractions of pulping liquors to produce fiber, energy, chemicals, and materials. A lignocellulosic-based biorefining strategy may be supported

Table 7.4 Common process units found in an oil refinery

	Unit	Treatment
1	Desalter	Washes out salt from the crude oil before it enters the atmospheric distillation unit
2	Atmospheric distillation	IDdistills crude oil into fractions
3	Vacuum distillation	Further distills residual bottoms after atmospheric distillation
4	Naphtha hydrotreater	Uses hydrogen to desulfurize naphtha from atmospheric distillation
5	Catalytic reformer	Used to convert the naphtha-boiling-range molecules into higher-octane reformer product (reformate)
6	Distillate hydrotreater	Desulfurizes distillates (such as diesel) after atmospheric distillation
7	Fluid catalytic cracker (FCC)	Upgrades heavier fractions into lighter, more valuable products
8	Hydrocracker	Uses hydrogen to upgrade heavier fractions into lighter, more valuable products
9	Visbreaking	Upgrades heavy residual oils by thermally cracking them into lighter, more valuable reduced-viscosity products
10	Merox	Treats LPG, kerosene, or jet fuel by oxidizing mercaptans into organic disulfides
11	Coking	Converts very heavy residual oils into gasoline and diesel fuel, leaving petroleum coke as a residual product
12	Alkylation	Produces high-octane component for gasoline blending
13	Dimerization	Converts olefins into higher-octane gasoline blending components
14	Isomerization	Converts linear molecules into higher-octane branched molecules for blending into gasoline
15	Steam reforming	Produces hydrogen for hydrotreaters or hydrocracker
16	Liquefied gas storage	For propane and similar gaseous fuels at pressure sufficient to maintain in liquid form
17	Storage tanks	For crude oil and finished products, usually cylindrical, with some sort of vapor emission control
18	Amine gas and tail gas treatment	For converting hydrogen sulfide from hydrodesulfurization into elemental sulfur
19	Utility units	Such as cooling towers for circulating cooling water, boiler plants for steam generation
20	Wastewater collection and treating	Converts wastewater into water suitable for reuse or for disposal
21	Solvent refining	Uses solvent such as cresol or furfural to remove unwanted, mainly asphaltenic materials from lubricating oil stock
22	Solvent dewaxing	For removing the heavy waxy constituents of petroleum from vacuum distillation products

7.4 Petroleum Refinery and Biorefinery 173

Table 7.5 Classification of biorefineries based on their feedstocks

	Feedstocks	Products
Green biorefinery	Grasses, green plants	Ethanol
Cereal biorefinery	Starch crops, sugar crops, grains	Bioethanol
Oilseed biorefinery	Oilseed crops, oil plants	Vegetable oils, biodiesels
Forest biorefinery	Forest harvesting residues, barks, sawdust, pulping liquors, fibers	Fuels, energy, chemicals, materials
Lignocellulosic biorefinery	Agricultural wastes, crop residues, urban wood wastes, industrial organic wastes	Lignocellulosic ethanol, bio-oil, gaseous products

by biomass reserves, created initially with residues from wood product processing or agriculture. Biomass reserves should be used to support first-generation biorefining installations for bioethanol production, development of which will lead to the creation of future high-value coproducts (Mabee *et al.* 2006).

Biorefineries can be classified based on their production technologies: first-generation biorefineries (FGBRs), second-generation biorefineries (SGBRs), third-generation biorefineries (TGBRs), and fourth-generation biorefineries.

The FGBRs refer to biofuels made from sugar, starch, vegetable oils, or animal fats using conventional technology. Table 7.6 shows the classification of biorefineries based on their generation technologies. SGBRs and TGBRs are also called advanced biorefineries. SGBRs made from nonfood crops, wheat straw, corn, wood, and energy crop using advanced technology.

Sugar and vegetables are used and converted into bioalcohols and biodiesel in FGBRs. The transition from FGBRs to SGBRs will mark a qualitative leap. Lignocellulosic residues such as sugar cane bagasse and rice straw feedstocks are used and converted into SGBs in SGBRs.

The first TGBR demonstration plant in the world was commissioned in Oulu, Finland, by Chempolis Oy. As far as is known, the world's first TGBR producing paper fiber, biofuel, and biochemicals from nonwood and nonfood materials was launched in Finland. TGBRs start with a mix of biomass feedstocks (agricultural or forest biomass) and produce a multiplicity of various products, such as ethanol for fuels, chemicals, and plastics, by applying a mix of different (both small- and

Table 7.6 Classification of biorefineries based on their generation technologies

Generation	Feedstocks	Examples
First	Sugar, starch, vegetable oils, animal fats	Bioalcohols, vegetable oil, biodiesel, biosyngas, biogas
Second	Non food crops, Wheat straw, Corn, Wood, Solid waste, Energy crop	Bioalcohols, bio-oil, bio-DMF, bio-hydrogen, bio-Fischer–Tropsch diesel
Third	Algae	Vegetable oil, biodiesel, jet fuel
Fourth	Vegetable oil, biodiesel	Biogasoline

large-scale) technologies such as extraction, separation, and thermochemical or biochemical conversion. However, large integrated TGBRs are not expected to become fully established until around 2020. Increasing quantities of agricultural residues will be needed to make paper in the future, as insufficient wood is available locally in the world's growing paper markets, forest resources are declining, and growing environmental pressures are being put on the use of wood. Vegetable oil is used and converted into biogasoline in fourth-generation biorefineries.

Biorefineries can also be classified based on their conversion routes: biosyngas-based biorefineries, pyrolysis-based biorefineries, hydrothermal-upgrading-based biorefineries, fermentation-based biorefineries, and oil-plant-based biorefineries. Table 7.7 shows the classification of biorefineries based on their conversion routes. Biosyngas is a multifunctional intermediate for the production of materials, chemicals, transportation fuels, power, and heat from biomass.

Table 7.7 Classification of biorefineries based on their conversion routes

Biorefinery	Products
Biosyngas-based	Syngas, hydrogen, methanol, dimethyl ether, FT diesel
Pyrolysis-based	Bio-oil, diesel fuel, chemicals, oxygenates, hydrogen
Hydrothermal-upgrading-based	C_xH_x, diesel fuel, chemicals
Fermentation-based	Bioethanol
Oil-plant-based	Biodiesel, diesel fuel, gasoline

7.5 Refining of Upgraded Products

A biorefinery system starts with the contract harvesting of whole crops (grain and straw), which are then stored and fractionated (including drying as necessary) into products and byproduct for sale. A biorefinery is a factory that processes crops, such as wheat, barley, and oilseed rape, to produce various refined specialized fractions, such as flour, gluten, starch, oil, straw chips, etc. The concept of a biorefinery, compared with, for example, a flour mill, is that the use and value of all the fractions into which the input can be separated is maximized (Audsley and Annetts 2003).

The analysis of a biorefinery system can be considered in three parts. The first is the effect on the farm of selling products to a biorefinery, on the assumption that the biorefinery contracts to harvest the crop using a wholecrop forage harvester. The second part is the impact of the type of biorefinery system on the profitability of the processing required to produce the various products. The third part is the transport of crop to the biorefinery, which is a function of the distribution of farms around the biorefinery location (Audsley and Annetts 2003).

As biomass hydrolysis and sugar fermentation technologies approach commercial viability, advancements in product recovery technologies will be required. For cases in which fermentation products are more volatile than water, recovery by distillation is often the technology of choice. Distillation technologies that will allow

7.5 Refining of Upgraded Products

the economic recovery of dilute volatile products from streams containing a variety of impurities have been developed and commercially demonstrated. A distillation system separates the bioethanol from water in the liquid mixture.

The first step is to recover the bioethanol in a distillation or beer column, where most of the water remains with the solids part. The product (37% bioethanol) is then concentrated in a rectifying column to a concentration just below the azeotrope (95%). The remaining bottom product is fed to the stripping column to remove additional water, with the bioethanol distillate from stripping being recombined with the feed to the rectifier. The recovery of bioethanol in the distillation columns in the plant is fixed at 99.6% to reduce bioethanol losses.

After the first effect, solids are separated using a centrifuge and dried in a rotary dryer. A portion (25%) of the centrifuge effluent is recycled to fermentation and the rest is sent to the second and third evaporator effects. Most of the evaporator condensate is returned to the process as fairly clean condensate (a small portion, 10%, is split off to wastewater treatment to prevent buildup of low-boiling-point compounds) and the concentrated syrup contains 15 to 20% by weight total solids.

One of the advantages of alkane production from biomass by aqueous-phase dehydration/hydrogenation is that the alkanes spontaneously separate from the aqueous feed solution, whereas ethanol produced during fermentation processes must be removed from water by an energy-intensive distillation step.

Biomass-derived oxygenates can be converted into hydrogen and alkanes (ranging from C_1 to C_{15}) via aqueous-phase processing (Audsley and Annetts 2003). These aqueous-phase processes could be used in an integrated biorefinery to produce a range of fuels, as shown in Figure 7.5. The first step in the biorefining process is conversion of biomass into an aqueous sugar solution. Production of hydrogen for biorefining processes is accomplished by aqueous-phase reforming. The biorefinery can also produce light alkanes ranging from C_1 to C_6 by aqueous-phase dehydration/hydrogenation (Audsley and Annetts 2003). The light alkanes could be used as synthetic natural gas, liquefied petroleum gas, and a light naptha stream. Aqueous-phase processing can also produce larger alkanes ranging from C_7 to C_{15} by combining the dehydration/hydrogenation reactions with an aldol condensation step prior to the aqueous-phase dehydration/hydrogenation step (Huber *et al.* 2005).

Biomass has been traditionally converted into liquid fuels by either (a) fermentation or (b) pyrolysis methods. Modern improvements to these classical processes are many in number but do not essentially change the type of product resulting from these two vastly different sets of reaction conditions. While ethanol production by fermentation has become more efficient, it is still limited to a 67% yield due to the loss of one third (1/3) of the available carbon as carbon dioxide gas. Pyrolytic reactions also lose carbon as gases and char but may achieve about 80% carbon conversion. While most thermochemical processes usually require nearly dry feedstock, the hydrothermal upgrading (HTU) process requires a 3:1 ratio of water to biomass. However, HTU produces only 50% biocrude, which still contains 10 to 15% oxygen. Obviously, there remains a need for a variety of fuels from many sources, especially conventional liquid fuels for transportation purposes. To resolve this fuel problem

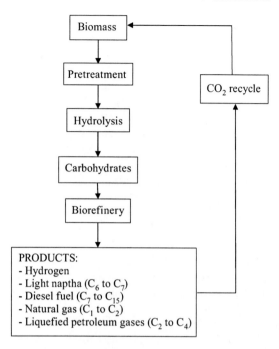

Figure 7.5 An integrated biorefinery for conversion of carbohydrates into fuels by aqueous-phase processing

and to use a renewable resource, a strategy was selected to prepare valuable liquid hydrocarbons from biomass by a new chemical process.

Glycerol can be converted into higher-value products. The products are 1,3-propanediol, 1,2-propanediol, dihydroxyacetones, hydrogen, polyglycerols, succinic acid, and polyesters. The main glycerol-based oxygenates are 1,3-propanediol, 1,2-propanediol, propanol, glycerol tertbutyl ethers, ethylene glycol, and propylene glycol. Glycerol has been pyrolyzed for the production of clean fuels such as H_2 or a feedstock such as syngas for additional transportation fuel via FTS. The conversion of glycerol to H_2 and CO takes place according to the following stoichiometric equation:

$$C_3O_3H_8 \rightarrow 3CO + 4H_2 \tag{7.1}$$

The stoichiometry for conversion of glycerol into liquid alkanes, by the formation of synthesis gas coupled with FTS, is shown in Equation 7.2:

$$25C_3O_3H_8 \rightarrow 7C_8H_{18} + 19CO_2 + 37H_2O \tag{7.2}$$

This overall reaction to produce liquid fuels from glycerol is slightly exothermic and the yield of liquid alkanes is 40% at 1.7 MPa pressure.

It is possible to produce light alkanes by aqueous-phase reforming of biomass-derived oxygenates such as sorbitol, which can be obtained from glucose by hydrogenation (Huber et al. 2005; Metzger 2006). The production of alkanes from aqueous carbohydrate solutions would be advantageous because of the easy separa-

7.5 Refining of Upgraded Products

tion of the alkanes from water. Much hydrogen is needed to reduce biomass-derived oxygenates to alkanes as shown in Equation 7.3:

$$C_6H_{14}O_6 + 6H_2 \rightarrow C_6H_{14} + 6H_2O \qquad (7.3)$$

Production of ethanol (bioethanol) from biomass is one way to reduce both the consumption of crude oil and environmental pollution. Ethanol from lignocellulosic biomass has the potential to contribute substantially to bioethanol for transportation. In the process evaluated, prehydrolysis with dilute sulfuric acid is employed to hydrolyze hemicellulose and make the cellulose more accessible to hydrolysis by enzymes. Residual biomass from hydrolysis and extraction of carbohydrates can be burned in a power plant to generate electricity and process steam. Figure 7.6 shows a flow diagram of pretreatment for fermentation of ethanol production from sugar crops and lignocellulosic feedstocks.

Carbohydrates (hemicelluloses and cellulose) in plant materials can be converted into sugars by hydrolysis. Fermentation is an anaerobic biological process in which sugars are converted into alcohol by the action of microorganisms, usually yeast. The resulting alcohol is bioethanol. The value of any particular type of biomass as feedstock for fermentation depends on the ease with which it can be converted into sugars. Bioethanol is a petrol additive/substitute. Bioethanol and the biorefinery concept are closely linked. It is possible that wood, straw, and even household wastes may be economically converted into bioethanol. In 2004, 3.4 billion gal. of fuel

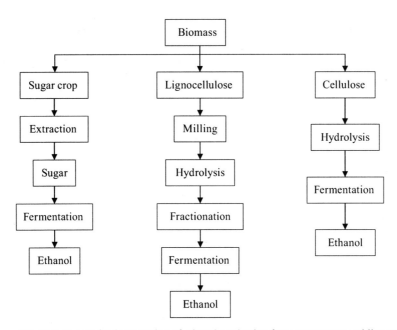

Figure 7.6 Pretreatment for fermentation of ethanol production from sugar crops and lignocellulosics

ethanol were produced from over 10% of the corn crop. Ethanol demand is expected to more than double in the next 10 years. For the supply to be available to meet this demand, new technologies must be moved from the laboratory to commercial reality. World ethanol production is about 60% from sugar-crop feedstock.

The corn-starch-to-fuel-ethanol industry has been developed over the past 30 years by bioethanol researchers. Most bioethanol researchers focus on the challenge of producing bioethanol from lignocellulosic biomass instead of from corn starch. To this end, researchers have already developed effective technology to thermochemically pretreat biomass, to hydrolyze hemicellulose to break it down into its component sugars and open up the cellulose to treatment, to enzymatically hydrolyze cellulose to break it down into sugars, and to ferment both 5-carbon sugars from hemicellulose and 6-carbon sugars from cellulose.

Cellulose is a remarkable pure organic polymer, consisting solely of units of anhydroglucose held together in a giant straight-chain molecule. Cellulose must be hydrolyzed to glucose before fermentation into ethanol. Conversion efficiencies of cellulose to glucose may depend on the extent of chemical and mechanical pretreatments to structurally and chemically alter the pulp and paper mill wastes. The method of pulping, the type of wood, and the use of recycled pulp and paper products also could influence the accessibility of cellulose to cellulase enzymes.

Cellulose fraction of the structural components is insoluble in most solvents and has a low accessibility to acid and enzymatic hydrolysis. Hemicelluloses (arabinoglycuronoxylan and galactoglucomammans) are related to plant gums in composition and occur in much shorter molecule chains than cellulose. The hemicelluloses, which are present in deciduous woods chiefly as pentosans and in coniferous woods almost entirely as hexosanes, undergo thermal decomposition very readily. Hemicelluloses are derived mainly from chains of pentose sugars and act as the cement material holding together the cellulose micells and fiber. Hemicelluloses are largely soluble in alkali and as such are more easily hydrolyzed.

7.6 Opportunities for Refining Pyrolysis Products

Biorefinery refers to the conversion of biomass feedstock into value-added chemicals and fuels with minimal waste and emissions. Some current aspects of biorefinery research and development since the early 1990s are presented, revealing that integrated processes, biomass upgrading technology, and biorefinery technology have become objects of research and development. Many of the currently used biorenewable-based industry products are the results of direct physical, catalytic, or chemical treatment and processing of biomass feedstocks.

The primary objective refining of pyrolysis products is to obtain valuable fuels like gasoline, diesel and jet fuel, and chemicals from biomass. Figure 7.7 shows the products from biomass by pyrolysis-based refining.

In vacuum pyrolysis, biomass is heated in a vacuum in order to decrease the boiling point and avoid adverse chemical reactions. In flash vacuum thermolysis

7.6 Opportunities for Refining Pyrolysis Products

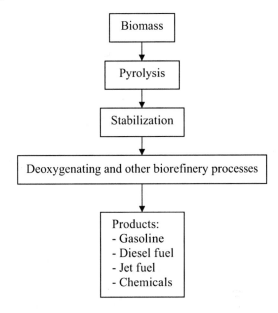

Figure 7.7 Products from biomass by pyrolysis-based refinery

(FVT), the residence time of the substrate at the working temperature is limited as much as possible, again in order to minimize secondary reactions.

Vacuum pyrolysis is the thermal degradation of a feedstock in the absence of oxygen and under low pressure to produce a bio-oil and char as main products, together with water and noncondensable gases. Both the bio-oil and char have a high energy content and may be used as fuels. An incredible number of chemical compounds are also found in the bio-oil, and these compounds can be extracted and sold as high-value chemicals.

Vacuum pyrolysis is a relatively new variant of pyrolysis with many recycling applications. During vacuum pyrolysis of biomass, the biorenewable feedstock is thermally decomposed under reduced pressure. The quick removal of the vapors reduces the residence times of the macromolecules and hence minimizes secondary decomposition reactions such as cracking, repolymerization, and recondensation, which occur during atmospheric pyrolysis. Temperatures between 675 K and 775 K and pressures of about 0.15 atm are typically used. Table 7.8 compares pyrolysis and gasification processes.

Bio-oil contains the thermally cracked products of the original cellulose, hemicelluloses, and lignin fractions present in biomass. It also contains a high percentage of water, often as high as 30%. The total oil is often homogeneous after quenching but can easily be separated into two fractions, a water-soluble fraction and a heavier pyrolytic lignin fraction. The addition of more water allows the pyrolytic lignin fraction to be isolated, and the majority of it consists of the same phenolic polymer as lignin but with smaller-molecular-weight fragments. Bio-oils are composed of a range of cyclopentanone, methoxyphenol, acetic acid, methanol, acetone, furfural, phenol, formic acid, levoglucosan, guaioco,l and their alkylated phenol derivatives.

Table 7.8 Comparison of pyrolysis and gasification processes

		Products, wt%		
Process	Conditions	Liquid	Char	Gas
Fast pyrolysis	Moderate temperature (700–800 K)			
	Short hot vapor residence time (< 2 s)	70–77	10–14	11–15
Slow pyrolysis	Low-moderate temperature (600–725 K)			
	Long residence time	27–33	33–38	33–38
Vacuum pyrolysis	Moderate temperature (675–775 K)			
	Low pressure (< 0.15 atm)	74–80	8–12	9–14
Gasification	High temperature (> 1,075 K)			
	Long residence time	4–6	8–12	82–88

Pyrolytic lignin is a better feedstock for liquid fuel production than the water-soluble fraction because of its lower oxygen content, and therefore the study focused on evaluating it as a potential feedstock for the production of highly aromatic gasoline (Demirbas 2000).

The pyrolysis of biomass is a thermal treatment that results in the production of charcoal, liquid, and gaseous products. Among the liquid products, methanol is one of the most valuable. The liquid fraction of the pyrolysis products consists of two phases: an aqueous phase containing a wide variety of organooxygen compounds of low molecular weight and a nonaqueous phase containing insoluble organics of high molecular weight. This phase is called tar and is the product of greatest interest. The ratios of acetic acid, methanol, and acetone of the aqueous phase are higher than those of the nonaqueous phase.

The bio-oil formed at 725 K contain high concentrations of compounds such as acetic acid, 1-hydroxy-2-butanone, 1-hydroxy-2-propanone, methanol, 2,6-dimethoxyphenol, 4-methyl-2,6-dimetoxyphenol, 2-cyclopenten-1-one, etc. A significant characteristic of bio-oils is the high percentage of alkylated compounds, especially methyl derivatives (Demirbas 2007).

References

Audsley, E., Annetts, J. E. 2003. Modelling the value of a rural biorefinery – part I: The model description. Agricult Sys 76:39–59.

Balat, M. 2007. Production of biodiesel from vegetable oils: a survey. Energy Sources A 29:895–913.

Balat, M. 2009. New biofuel production technologies. Energy Educ Sci Technol A 22:147–161.

Chhetri, A. B., Islam, M. R. 2008. Towards producing a truly green biodiesel. Energy Sources A 30:754–764.

Demirbas, A. 2000. Mechanisms of liquefaction and pyrolysis reactions of biomass. Energy Convers Manage 41:633–46.

Demirbas, A. 2002. Biodiesel from vegetable oils via transesterification in supercritical methanol. Energy Convers Manage 43:2349–2356.

References

Demirbas, A. 2003. Biodiesel fuels from vegetable oils via catalytic and non-catalytic supercritical alcohol transesterifications and other methods: a survey. Energy Convers Manage 44:2093–2109.

Demirbas, A. 2007. Converting biomass derived synthetic gas to fuels via Fisher–Tropsch synthesis. Energy Sources A 29:507–512.

Demirbas, A. 2008. Recent progress in biorenewable feedstocks. Energy Educ Sci Technol 22:69–95.

Demirbas, M.F. 2009. Biorefineries for biofuel upgrading: a critical review. Appl Energy 86 (Suppl 1):S151–S161.

Demirbas, A., Karslioglu, S. 2007. Biodiesel production facilities from vegetable oils and animal fats. Energy Sources A 29:133–141.

Dincer, K. 2008. Lower emissions from biodiesel combustion. Energy Sources Part A 30:963–968.

GBRP. 2007. Review of current state of biorefinery development in G8+5. Global Biorefinery Partnership, New York.

Goldemberg, J. 2000. World Energy Assessment. United Nations Development Programme, New York.

Huber, G. W., Chheda, J. N., Barrett, J. A. 2005. Dumesic, production of liquid alkanes by aqueous-phase processing of biomass-derived carbohydrates. Science 308:1446–1450.

Kamm, B., Kamm, M., Gruber, P. (eds.). 2005. Biorefineries – Biobased industrial processes and products. status quo and future directions, WILEY–VCH, Weinheim.

Keskin, A. 2009. Biodiesel production from free fatty acids obtained with neutralization of the crude glycerin. Energy Sources A 31:17–24.

Khoiyangbam, R. S. 2008. Greenhouse gases emission from energy production in conventional biogas plants in India. Energy Sources A 30:689–697.

Mabee, W. E., Fraser, E. D. G., McFarlane, P. N., Saddler, J. N. 2006. Canadian biomass reserves for biorefining. Appl Biochem Biotechnol 129:22–40.

Metzger, J. O. 2006. Production of liquid hydrocarbons from biomass. Angew Chem Int Ed 45:696–698.

Nova Scotia Museum. 2008. Abraham Gesner. Fossils of Nova Scotia. Web site: http://museum.gov.ns.ca/fossils/finders/gesner.htm.

Parikka, M. 2004. Global biomass fuel resources. Biomass Bioenergy 27:613–620.

WI (Worldwatch Institute). 2007. Biofuels for Transport: Global Potential and Implications for Energy and Agriculture, Earthscan, London.

Chapter 8
Future Developments

8.1 Introduction

During the last 200 years, developed countries have shifted their energy consumption toward fossil fuels. Renewable energies have been the primary energy source in the history of the human race. Wood was used for cooking and water and space heating. The first renewable energy technologies were primarily simple mechanical applications and did not reach high energetic efficiencies. Industrialization changed the primary energy use from renewable resources to sources with a much higher energetic value such as coal and oil. The promise of unlimited fossil fuels was much more attractive, and rapid technical progress made the industrial use of oil and coal economical.

Developing renewable sources of energy has become necessary due to the limited supply of fossil fuels. Global environmental concerns and decreasing resources of crude oil have prompted demand for alternative fuels. Global climate change is also the major environmental issue of current times. Global warming, the Kyoto Protocol, the emission of greenhouse gases, and the depletion of fossil fuels are the topics of environmental pleadings worldwide. Due to rapidly increasing energy requirements along with technological development around the world, research and development activities have perforce focused on new and renewable energy.

Competition of renewable liquid fuels into petroleum liquid fuels will be important in the near future. This can be achieved by research and development (R&D), technological development, and industrial mobilization by implementing a proper energy tax system that takes into account the environmental and social costs of conventional energies, by making calculations based on the entire energy system and not only the cost of one technology but of the entire energy chain, by calculating the burdens for the national economy of every country from importing fossil energies, etc. This requires a comprehensive view on energy.

The calculation of energy prices should be designed to maintain into the future the equilibrium between demand and supply, taking into account the costs of planned investments. They should also take into account the rest of the economy and

A. Demirbas, M. Fatih Demirbas, *Algae Energy*
DOI 10.1007/978-1-84996-050-2, © Springer 2010

the environment. Two very important characteristics of energy prices are equity and affordability. Energy prices must reflect the cost imposed by the specific consumer category on the economy. Since energy prices based on apparent long-run marginal costs may not be sufficient to finance the development of the energy sector, prices should be adjusted so that the energy sector can be financed without subsidies to enhance its autonomy. In competitive markets this form of adjustment may not be possible. Energy pricing policy should not be employed as an anti-inflationary instrument. It should be applied in such a way that it does not create cross-subsidies between classes of consumers.

8.1.1 World Theoretical Limit of Biomass Supply

The amount of biomass available is limited because plants, on average, capture only about 0.1% of the solar energy reaching the Earth (Pimentel and Pimentel 1996). Temperature, water availability, soil nutrients, and the feeding pressure of herbivores all limit biomass production in any given region. Under optimal growing conditions, natural and agricultural vegetation produce about 12 million kcal/ha/year (approx. 3 t/ha dry biomass). The productive ecosystems in the world total an estimated 50 billion ha, excluding the icecaps. Marine ecosystems occupy approx. 36.5 billion ha, while the terrestrial ecosystems occupy approx. 13.5 billion ha.

Sustainable production of biomass will limit supply. The total biomass produced is approx. 77 billion t or approx. 12.6 t/person/year (Pimentel 2001). Globally, suitable abandoned cropland and pastureland amounts to approx. 1.5 million square miles. Realistically, energy crops raised on this land could be expected to yield about 27 exajoules (EJ) of energy each year (1 EJ $= 10^{18}$ J). This is a huge amount of energy, equivalent to 172 million barrels of oil. In 2003 the EU biomass production was about 69 Mtoe, covering about 4% of EU energy needs; the production potential is estimated to increase to 186 to 189 Mtoe in 2010, to 215 to 239 Mtoe in 2020, and up to 243 to 316 Mtoe in 2030 (EEA 2005). One analysis carried out by the UN Conference on Environment and Development (UNCED) estimates that biomass could potentially supply about half of the present world primary energy consumption by the year 2050 (Ramage and Scurlock 1996).

Global biomass production on the Earth's land surface is equal to 4,560 EJ (the gross primary production), of which half is lost by autotrophic respiration and decomposition, leaving 2,280 EJ (net primary production or NPP) (Smeets *et al.* 2007). The availability of the NPP for use in food and energy production is restricted by many factors, e.g., logistics, economics, or legal restraints. Without intervention this NPP is in balance with natural decomposition. There are three types of biomass energy sources: dedicated bioenergy crops, agricultural and forestry residues and waste, and forest growth. The bioenergy potential in a region is limited by various factors, such as the demand for food, industrial round wood, traditional wood fuel, and the need to maintain existing forests for the protection of biodiversity. The global potential of bioenergy production from agricultural and forestry residues and

8.1 Introduction 185

wastes was calculated to be 76 to 96 EJ/year by the year 2050. The potential of bioenergy production from surplus forest growth was calculated to be 74 EJ/yr by 2050 (Smeets *et al.* 2007).

Biomass resources can be divided into six categories: energy crops on surplus cropland, energy crops on degraded land, agricultural residues, forest residues, animal manure, and organic wastes. The range of the global potential of primary biomass (in about 50 years) is very broadly quantified at 33 to 1,135 EJ/year (Hoogwijk *et al.* 2003).

Plant height, main stem diameter, stems, leaves, leaf length, leaflet width and length, and leaflets are important traits that are used to estimate herbage yield (Ates and Tekeli 2005). Trait characterization is part of sustainable crop systems. The improvement of crops for tolerance to various forms of abiotic stress and for utilization in semiarid regions can be achieved by using trait analyses in multiple environments. This involves analyzing crop phenotypes for stress physiology and agronomic traits (e.g., high yield, grain quality) in different locations under different growth environments.

There is a need to accelerate breeding applications to improve quality traits in crops that contribute to food security, health, and agricultural sustainability. The complex genetics and quality traits of many crops are difficult to manipulate by conventional breeding. There is a lack of useful variability for key quality traits and stress tolerance in cultivars and adapted germplasm.

8.1.2 High-yield Energy Crops

The main energy crops are short-rotation woody crops, herbaceous woody crops, grasses, starch crops, sugar crops, forage crops, and oilseed crops. Energy crops are fast-growing, genetically improved trees and grasses grown under sustainable conditions for harvest at 1 to 10 years of age.

Agricultural residues, grasses, algae, kelps, lichens, and mosses are also important biomass feedstocks in the world. Algae can grow practically wherever there is sufficient sunshine. Some algae can grow in saline water. The most significant feature of algal oil is its yield and, hence, its biodiesel yield. According to some estimates, the yield (per acre) of oil from algae is over 200 times the yield from the best-performing plant/vegetable oils (Sheehan *et al.* 1998). Microalgae are the fastest growing photosynthesizing organisms. They can complete an entire growing cycle every few days. Approximately 46 tons of oil/ha/year can be produced from diatom algae. Different algae species produce different amounts of oil. Some algae produce up to 50% oil by weight.

Commercial energy crops are typically densely planted, high-yielding crop species where the energy crops are burnt to generate power. Woody crops such as willow and poplar are widely utilized, and tropical grasses such as miscanthus and pennisetum purperium (both known as elephant grass) are receiving more attention from emerging energy crop companies.

186   8 Future Developments

Genetic research into dedicated energy crops is still at a very early stage. Current research is focused on mapping gene sequences and identifying key locations where modifying genetic code could provide significant benefits. Modern biotechnology can also be used for increasing yields and modifying plant characteristics to enhance their conversion into fuels. Genetic engineering may result in energy crops that have a higher percentage of cellulose or hemicellulose and lower lignin content for increasing alcohol production yield, as well as a greater ability to take up carbon in their root systems. Crops could also be modified to produce large quantities of the enzymes that are necessary for feedstock conversion into ethanol. Oilseed crops could be bioengineered to become the source of bio-based lubricants and esterified fatty acids, which are the main ingredient in biodiesel (UN 2006).

Genetic modification of fuel-dedicated crops may raise fears linked to perceived threats of agrobiotechnology to plant life and health, to the conservation of biodiversity, and to the environment at large. The environmental, sustainability, and public-perception aspects of genetically modified energy crop plantations should be carefully evaluated before widespread production starts (UN 2006).

8.1.3 Food Versus Fuel Delineation

Serious problems face the world food supply today. The rapidly growing world population and rising consumption of fossil fuels is increasing demand for both food and biofuels. This will exacerbate both food and fuel shortages. The human population faces serious food shortages and malnutrition (WHO 2005).

Producing biofuels requires huge amounts of both fossil energy and food resources, which will intensify conflicts among these resources. Using food crops such as corn grain to produce ethanol raises major nutritional and ethical concerns. Nearly 60% of humans in the world are currently malnourished, so the need for grains and other basic foods is critical (WHO 2005). Growing crops for fuel squanders land, water, and energy resources vital for the production of food for people.

Food versus fuel is the dilemma regarding the risk of diverting farmland or crops for liquid biofuel production to the detriment of the food supply on a global scale. There is disagreement about how significant this is, what is causing it, what the impact is, and what can or should be done about it. Biofuel production has increased in recent years. Some commodities such as corn, sugar cane, and vegetable oil can be used as food or feed or to make biofuels. For example, vegetable oils have become more attractive recently because of their environmental benefits and the fact that they are made from renewable resources. Vegetable oils are a renewable and potentially inexhaustible source of energy, with energy content close to that of diesel fuel. On the other hand, extensive use of vegetable oils may cause other significant problems such as starvation in developing countries.

Several studies have shown that biofuel production can be significantly increased without increased acreage.

8.1.4 Thermodynamic Efficiency (Exergy Analysis)

Exergy is defined as the maximum amount of work that can be obtained from a material stream, heat stream, or work interaction by bringing this stream to environmental conditions. When the surroundings are the reservoir, exergy is the potential of a system to cause a change as it achieves equilibrium with its environment. Exergy is then the energy that is available to be used. After the system and surroundings reach equilibrium, the exergy is zero. Energy is never destroyed during a process; it changes from one form to another.

Exergy analysis is a relatively new method of thermodynamic analysis that has recently been applied in power and heat technology, chemical technology, and other fields of engineering and science. The exergy method takes into account not only the quantity of materials and energy flows but also the quality of materials and energy flows. The exergy concept is based on both the first and second laws of thermodynamics. The main reason for exergy analysis is to detect and evaluate quantitatively the losses that occur in thermal and chemical processes (Ptasinski 2008).

The development of efficient technologies for biomass gasification and synthesis of biofuels requires the correct use of thermodynamics. Among the different forms of exergy, three forms are the major contributors to total exergy: thermal exergy, work exergy, and exergy of material, which contains chemical and physical exergy components.

Exergy analysis is a convenient tool for the development and optimization of future biomass processes. Biomass gasification followed by synthesis of biofuels seems to be more promising for the medium and longer terms. The conversion efficiency of all investigated biomass-to-biofuel routes can be increased by improving the operation of biomass gasifiers, which show the highest exergy losses in all considered processes. The exergetic efficiency of biomass-to-biofuel processes depends not only on the feedstock quality but also on the degree of energy integration in these processes (Ptasinski 2008).

8.1.5 Biofuel Upgradation

The conversion of wood into chemicals for the production of most of our synthetic plastics, fibers, and rubbers is technically feasible. Synthetic oils from liquefaction of wood might serve as feedstocks for cracking into chemicals in the same way that crude oil is presently used.

Upgrading of condensed liquid from biomass involves three stages. There are physical upgrading (differential condensation, liquid filtration, and solvent addition), catalytic upgrading (deoxygenating and reforming), and chemical upgrading (new fuel and chemical synthesis).

The bio-oil obtained from the fast pyrolysis of biomass has high oxygen content. Because of the reactivity of oxygenated groups, the main problem of the bio-oil is its instability. Therefore, study of the deoxygenation of bio-oil is needed. In previous work the mechanism of hydrodeoxygenation (HDO) of bio-oil in the presence of a cobalt molybdate catalyst was studied (Zhang *et al.* 2003).

The main HDO reaction is represented in Equation 8.1:

$$-(CH_2O)- + H_2 \rightarrow -(CH_2)- + H_2O \tag{8.1}$$

This is the most important route of chemical upgrading. Reaction 8.1 has strong analogies with typical refinery hydrogenations like hydrodesulfurization and hydrodenitrification. In general, most of the HDO studies have been performed using existing hydrodesulfurization catalysts (NiMo and CoMo on suitable carriers). Such catalysts need activation using a suitable sulfur source, and this is a major drawback when using nearly sulfur-free resources like bio-oil.

8.1.6 Carbon Credits

Carbon credits are a tradable permit scheme under the United Nations Framework Convention for Climate Change (UNFCCC), which gives the owner the right to emit one metric ton of carbon dioxide equivalent. They provide an efficient mechanism to reduce greenhouse gas (GHG) emissions by monetizing the reduction in emissions.

Global warming is caused by the emission of GHGs that get trapped in the atmosphere. Table 8.1 shows the global warming (GW) potential of gases. The potent GHGs are carbon dioxide, methane, nitrous oxide, hydroflourocarbons, perflouro-carbons, and sulfur hexafluoride (Humbad *et al.* 2009).

$$\text{CERs awarded} =$$

$$\text{Tons of GHG reduced } x \text{ GW potential of the gas (metric tons of C)} \tag{8.2}$$

GW is an imminent catastrophe with irreversible consequences. The Kyoto Protocol was adopted in Kyoto, Japan on 11 December 1997 and entered into force on 16 February 2005. One hundred eighty countries have ratified the treaty to date. It aims to reduce GHG emissions by 5.2% against the 1990 levels over the 5-year period 2008–2012. Developed countries are categorized under Annex 1 countries and are legally bound by the protocol, while the developing nations, categorized as Non-Annex 1 countries, which ratify the protocol, are not legally bound by it. The

Table 8.1 Global warming potential of gases

Greenhouse gas	Global warming potential
Carbon dioxide	1
Methane	21
Nitrous oxide	310
Hydroflourocarbons	140–11,700
Perflourocarbons	7,000–9,200
Sulfur hexaflouride	23,900

Kyoto Protocol has three mechanisms: joint implementation (JI), a clean development mechanism (CDM), and international emission trading (IET).

The CDM mechanism allows Annex 1 countries to meet their reduction targets by implementing emission reduction projects in Non-Annex 1 developing nations. A certified emission reduction (CER) is a certificate given by the CDM board to projects in developing countries to certify that they have reduced GHG emissions by one metric ton of carbon dioxide equivalent per year. These CERs are bought by the Annex 1 countries to meet their emission reduction targets.

Under JI, an Annex 1 party may implement an emission reduction project or a project that enhances removal by sinks in another Annex 1 country. It can use the resulting emission reduction units (ERUs) for meeting its target. Under the IET mechanism, countries can trade their surplus credits in the international carbon credits market to those countries with quantified emission limitation and reduction commitments under the Kyoto Protocol.

8.2 Social and Political Issues

Energy is an essential input for social development and economic growth. The role of energy as an essential catalyst to economic growth and an enhanced standard of living is a reality that policymakers should factor into their decisions and when evaluating environmental, economic, and social goals. The International Energy Agency estimates that world energy demand will increase by half again between now and 2030, with more than two thirds of this increase coming from developing and emerging countries. The population of developing countries is estimated to double by 2055, while the population of the industrial countries will increase by only 15% over the same period. New conventional fuel explorations, energy wars, and political maneuvers will not prevent the production of nonconventional fuels and the continuing evolution of a truly global energy market.

There are several reasons why biofuels are considered relevant technologies by both developing and industrialized countries. They include energy security reasons, environmental concerns, foreign exchange savings, and socioeconomic issues related to the rural sectors of all countries in the world. Biofuels could be peaceful energy carriers for all countries.

Bioenergy offers opportunities for additional value to be derived from products already in the economy. The dispersed nature of most biomass resources lends itself to smaller-scale operations of up to 50 MW. These are within the capability of communities to feed and operate, creating and retaining wealth within the local economy.

New employment opportunities arise in growing and harvesting biomass, transport and handling, and plant operation. They also extend to equipment manufacturers and maintenance crews. Farmers may improve returns as marginal crops become viable, giving an additional source of income from energy byproducts. Degraded forests may be rejuvenated and waste streams diverted to produce energy.

Bioenergy can also contribute to local and national energy security, which may be required to establish new industries. Bioenergy contributes to all important elements of national/regional development: economic growth through business earnings and employment, import substitution with direct and indirect effects on GDP and trade balance, and security and diversification of energy supply. Other benefits include support of traditional industries, rural diversification, and the economic development of rural societies. These are all important elements of sustainable development.

At the same time, global population and affluence growth have caused upward pressure on food prices, which has led to food insecurity in the developing world. Conversion of corn into ethanol, or the use of arable land and fresh water for biofuel crops like corn, sugar cane, or jatropha, has exacerbated this situation and linked food and energy in a dangerous way.

To satisfy energy needs and avoid putting pressure on foodstocks, an alternative biofuel is needed. Water algae's superior yields, combined with its ability to grow without arable land or fresh water, mean that algae is a far better biofuel candidate to replace oil than any land crop. An algaecentric energy-independence strategy would provide greater energy security while promoting poor-country development and without threatening food security.

8.2.1 The Promise of Algae: Energy Security

The cost of producing microalgal biodiesel can be reduced substantially by using a biorefinery-based production strategy, improving the capabilities of microalgae through genetic engineering and advances in engineering of photobioreactors (Chisti 2007). Genetic and metabolic engineering are likely to have the greatest impact on improving the economics of microalgal diesel production (Dunahay et al. 1995).

Algae offer the greatest promise of energy security for three reasons. Algae yield more than ten times as much oil per unit area as palm oil, the most productive land-based biofuel crop. Algae's yield depends upon exactly what species is used and especially upon the oil content, by weight, of that species. Algae's greater yield comes from basic biological differences from land plants and could be improved with research over time.

Algal oil is unlikely to be suitable for a standard refinery, meaning that either new refineries will need to be constructed or existing ones would need to be retrofitted to handle the different types of oil. Better processes for the conversion of algae into oil and the conversion of algal oil into usable fuel would make algae a far more efficient fuel source. Algae are unlike corn or soybeans in that there are not separate sowing and harvesting seasons. Their great yield is due to algae's ability to reproduce far more quickly than any land plant, and productivity should remain consistent through all seasons.

Over 50 countries worldwide possess the appropriate temperature range (293 to 303 K) for year-round cultivation of algae across Central America, South America, Africa, and Southeast Asia and Oceania. Algal productivity depends upon a set of climatic conditions prevailing across most of the Earth's equatorial region, and a monopoly or oligopoly over algae is very difficult to envision. Both coastal and inland states have unique advantages for cultivation, as coastal states possess direct access to salt water for algal culture as well as bays and coasts where algae could be cultivated using no land at all. Inland states fitting this temperature range typically have desert climates ideal for high-yield culture.

Algae can be cultivated in almost any area as long as it is flat, hot, sunny, and has access to salt water. It requires no arable land and no fresh water, and so will not compete for agricultural resources that are often scarce in areas best suited for algal growth. A quick survey of states suitable for algal culture reveals that almost all lie within the developing world and include some of the poorest countries on Earth. Many of these states suffer from poverty at least in part because they contain large swaths of arid, unfarmable land. Algae thrive in land that is too dry and too hot for conventional crops and can be grown in these climates without displacing food plots. Algal culture represents a significant opportunity for these developing states for several reasons.

Algae could provide these states with a cash crop that does not compete with subsidized crops grown in Europe and the USA. Powerful domestic agricultural lobbies represent a significant obstacle to large-scale import of agricultural products, particularly from the developing world, but feel less threatened when the product in question is mainly produced elsewhere for reasons of climate. Algae can be grown in salt water, adding no additional pressure to rivers and lakes already overdrawn by freshwater irrigation and reducing the number of minerals and nutrients that need to be added to culture media for growth.

8.3 Environmental Impacts of Biomass Production

Most traditional biofuels, such as ethanol from corn, wheat, or sugar beets, and biodiesel from oil seeds, are produced from classic agricultural food crops that require high-quality agricultural land for growth. An important parameter for biomass energy is the impact of land use. This impact category describes the environmental impact resulting from land use for human activities. In particular, the land use category considers natural land as a resource and assumes that land occupation and management causes consumption of the resource. Natural land can be defined as land not damaged at the moment by human activities and the remaining natural land fraction under use. Land use is an impact indicator.

Land use impacts have been related to the area of land used, or physical land use. Specifically, they include impacts on biodiversity, biotic production potential (including soil fertility and use value of biodiversity), and ecological soil quality (including life support functions of soil other than biomass production potential). The use of land to produce biofuels in a well-established infrastructure has an impact on the environment.

The environmental impacts of irrigation are the changes in the quantity and quality of soil and water as a result of irrigation and the ensuing effects on natural and social conditions at the tail-end and downstream of the irrigation scheme. Irrigated agriculture depends on supplies from surface or ground water. Irrigation agriculture will be an essential component of any strategy to increase the global food supply. Improving the environmental performance of irrigation agriculture is important for its long-term sustainability. Irrigation projects and irrigated agriculture practices can impact the environment in a variety of ways. The environmental impact of irrigation systems depends on the nature of the water source, the quality of the water, and how it is delivered to the irrigated land. The management of water application systems as well as the suitability of related agronomic practices has a dramatic influence on the environmental impact of irrigated agriculture.

The ecosystem diversity of fuel applications confers many environmental, economic, and consumer benefits. The use of crop residues for bioenergy production must be critically assessed because of its positive impact on soil carbon sequestration, soil quality maintenance, and ecosystem functions (Lal 2005).

Energy crops include fast growing trees such as hybrid poplar, black locust, willow, and silver maple, in addition to annual crops such as corn and sweet sorghum and perennial grasses such as switchgrass. Bioenergy from biomass, both residues and energy crops, can be converted into modern energy carriers such as hydrogen, methanol, ethanol, and electricity.

The impact of energy cropping on habitat and ecosystem diversity depends not only on the previous land use and cultivation but also on the nature of the energy crop. The role of energy cropping can play in the context of sustainable development, such as via the use of bioethanol, and its potential impact on biodiversity is a subject that warrants more concerted research (Adsavakulchai *et al.* 2004).

Conflicts exist today in the use of land, water, energy, and other environmental resources required by both food and biofuel production. Bioenergy supplies can be divided into two broad categories: (a) organic municipal waste and residues from the food and materials sectors and (b) dedicated energy crop plantations. The term "food supply chain" refers to the strict correlation and the functional link existing between the agriculture sector and transformation industry.

Serious problems face the world food supply today. The rapidly growing world population and rising consumption of biofuels is increasing demand for both food and biofuels. This exacerbates both food and fuel shortages. Using food crops such as corn grain to produce ethanol raises major nutritional and ethical concerns. Growing crops for fuel squanders land, water, and energy resources vital for the production of food for human consumption. Using corn for ethanol increases the price of US beef, chicken, pork, eggs, breads, cereals, and milk by 10 to 30% (Demirbas 2006; Pimentel *et al.* 2008).

Many problems associated with biofuels have been ignored by scientists and policymakers. The environmental impacts of corn ethanol are enormous: (1) corn production causes more soil erosion than any other crop grown; (2) more than 1,700 gal. of water are required to produce 1 gal. of corn ethanol; (3) enormous quantities of carbon dioxide are produced during corn ethanol production by the large quantity of fossil energy used in production, during fermentation, and when the soil is tilled, leaving soil organic matter exposed and oxidized. In addition, the conversion of cropland for biofuel production contributes to the release of GHGs, all of which speeds global warming; (4) using corn for ethanol increases the price of other foods dependent on manufacturing or animal feeding with corn (Demirbas and Demirbas 2007; Pimentel *et al.* 2008).

The use of soybeans as a potential biofuel source puts cropland in competition with food production. On the other hand, extensive use of vegetable oils in biodiesel production may cause other significant problems such as starvation in developing countries (Demirbas and Demirbas 2007; Demirbas 2007).

References

Adsavakulchai, S., Minns, D., Chan, A. 2004. Assessing the interaction of vegetation diversity and landuse using remote sensing: an example in southeastern Ontario, Canada. Environ Inf Arch 2:499–508.

Ates, E., Tekeli, A. S. 2005. Heritability and variance components of some morphological and agronomic traits in alfalfa (*Medicago sativa* L.). Proc Pakistan Acad Sci 42:1–5.

Chisti, Y. 2007. Biodiesel from microalgae. Biotechnol Adv 25:294–306.

Demirbas, A. 2006. Energy priorities and new energy strategies. Energy Educ Sci Technol 16:53–109.

Demirbas, A. 2007. Importance of biodiese as transportation fuel. Energy Policy 35:4661–4670.

Demirbas, A. H., Demirbas, I. 2007. Importance of rural bioenergy for developing countries. Energy Convers Manage 48:2386–2398.

Dunahay, T. G., Jarvis, E. E., Dais, S. S., Roessler, P. G. 1996. Manipulation of microalgal lipid production using genetic engineering. Appl Biochem Biotechnol 57–58:223–231.

EEA. 2005. How much biomass can Europe use without harming the environment, briefing. European Environmental Agency, Brussels, Belgium.

Hoogwijk, M., Faaij, A., van den Broek, R., Berndes, G., Gielen, D., Turkenburg, W. 2003. Exploration of the ranges of the global potential of biomass for energy. Biomass Bioenergy 25:119–133.

Humbad, A., Kumar, S., Babu, B. V. 2009. Carbon credits for energy self sufficiency in rural India – a case study. Energy Educ Sci Technol A 22:187–197.

Lal, R. 2005. World crop residues production and implications of its use as a biofuel. Environ Int 31:575–584.

Pimentel, D., Pimentel, M. 1996. Food, Energy and Society. Colorado University Press, Boulder, CO.

Pimentel, D. 2001. Biomass utilization, limits of. In: Meyers, R. A. (ed.). Encyclopedia of Physical Science and Technology, 3rd edn. Academic, San Diego.

Pimentel, D., Marklein, A., Toth, M. A., Karpoff, M., Paul, G. S., McCormack, R., Kyriazis, J., Krueger, T. 2008. Biofuel impacts on world food supply: use of fossil fuel, land and water resources. Energies 1:41–78.

Ptasinski, K. J. 2008. Thermodynamic efficiency of biomass gasification and biofuels conversion. Biofuels Bioprod Bioref 2:239–253.

Ramage, J., Scurlock, J. 1996. Biomass. In: Boyle, G. (ed.). Renewable energy-power for a sustainable future. Oxford University Press, Oxford.

Sheehan, J., Dunahay, T., Benemann, J., Roessler, P. 1998. A look back at the U.S. Department of Energy's Aquatic Species Program – Biodiesel from Algae. National Renewable Energy Laboratory (NREL) Report: NREL/TP-580-24190. Golden, CO.Smeets, E. M. W., Faaij, A. P. C.,

Lewandowski, I. M., Turkenburg, W. C. 2007. A bottom-up assessment and review of global bio-energy potentials to 2050. Prog Energy Combust Sci 33:56–106.

UN. 2006. The emerging biofuels market: regulatory, trade and development implications. United Nations Conference on Trade and Development, New York and Geneva.

WHO. 2005. Malnutrition worldwide. Environmental Burden of Disease Series, No. 12, World Health Organization, Geneva, Switzerland.

Zhang, S. P., Yan, Y. J., Ren, J. W., Li, T. C. 2003. Study of hydrodeoxygenation of bio-oil from the fast pyrolysis of biomass. Energy Sourc 25:57–65.

Index

A

Acetic acid   61, 126, 179, 10
Acetone   36, 166, 179, 180
Algae   10, 30–32, 49, 52, 61, 75–92, 97–103, 119–126, 129, 132, 133, 139–154, 159, 173, 185, 190, 191
Algae technology   75
  Definition of Algal Terms   76
  Production Systems   77
    harvesting microalgae   79
    closed and hybrid systems   87–89
    open-pond systems   83–87
    photobioreactors   80–83
    Production costs   89–93
Algal oil   52, 78, 90, 101, 103, 139, 140, 145, 146, 148, 149, 152, 185, 190
  Biodiesel   103, 141, 143, 144, 145, 147, 150–153
    advantages   153
    disadvantages   153, 154
    economy   151, 153, 154
  Chemicals   143
  Fractionation   145, 146
  Macroalgal oil   145
  Microalgal oil   78, 152
    extraction   146
  Production   141, 151
    Cladophora fracta   145
    Chlorella protothecoides   145
    photobioreactor   152
  Transesterification   141, 145
Animal fats   51, 57, 70, 76, 100–102, 140, 173
  Fish oil   133, 149
  Grease   22, 24, 59, 110

  yellow   108
Tallow (beef and sheep)   57, 102, 107

B

Bioalcohols   101, 160, 173
  Butanol   49, 53, 159
    biobutanol   22
  Ethanol   21–24, 30, 33, 36, 37, 42–45, 49–57, 70–72, 98, 104–117, 109–113, 154, 159, 161, 165–167, 173, 177,178, 186, 190–193
  Methanol   24, 25, 30, 36, 37, 41, 49, 53, 54, 56, 57, 60, 61, 71, 72, 102, 106, 107, 113, 141, 144–147, 159, 174, 179, 180, 192
  Propanol   53, 70, 176
Biodiesel   19, 21–26, 29, 33, 34, 50, 51–53, 56–60, 70, 75–80, 89, 90, 93, 97–116, 119, 129, 131, 139, 140–154, 159–166, 173, 174, 186, 189, 190, 191
  Advantages of biodiesel from algae oil   154
  Disadvantages of biodiesel from algal oil   154
  Economy   151
  Transesterification   24, 25, 56, 70, 102, 103, 107, 110, 114, 140, 141, 144–146, 150, 165
    catalytic   56, 102
    continuous transesterification process   57
    from algal oil   144–146, 150
    with supercritical methanol   25, 145,
Biofuels   19–23, 29, 30, 33, 49–54, 56, 58, 60, 62, 64–66, 68, 70–72, 76, 80, 88, 90, 98–101, 103, 105–107, 109–117, 119, 121, 123, 125, 140, 159, 160, 162,

163, 166, 167, 173, 186, 187, 189, 191, 192

Bioethanol 19, 21–24, 29, 33, 34, 36, 43, 50–55, 76, 97, 100, 101, 104, 106, 108, 109, 111, 112, 114, 117, 148, 152, 160, 163, 164, 173–175, 177, 178, 192

from microalga 148,

Biogas 19, 22, 32, 51, 61–63, 101, 106, 119, 173

Biogasoline 52, 101, 173, 174

Biohydrogen 19, 21, 22, 49, 50, 69, 77, 98, 101, 114, 125, 126, 159, 160, 173

Biomethanol 19, 22, 24, 109, 112, 114

Bio-oil 39, 41, 45, 50, 51, 59, 61, 65, 77, 98, 101, 102, 117, 118, 120, 121, 123, 124, 146, 149, 160, 162, 163, 173, 174, 179, 180, 187, 188

Biosyngas 24, 36, 64, 65, 68, 71, 98, 101, 162, 173, 174

Production technologies 51, 100, 101, 106, 173,

Biomass 1, 2, 10, 15, 17, 19–24, 29–39, 41, 43–45, 49–55, 59–61, 63–68, 71, 76–82, 85–87, 89, 90, 92, 93, 97–104, 106, 111, 112, 115, 117–119, 125, 126, 128–132, 139–143, 146–148, 151, 153, 159, 160–166, 171, 173–185, 187, 189, 191, 192

Cellulose 23, 39, 42–45, 62, 104, 105, 128, 166, 168, 177–179, 186

Glucose 23, 39, 42–44, 78, 105, 126, 146, 176, 178

Hemicelluloses 23, 24, 36, 42–45, 62, 104, 105, 128, 177–179, 186

Landfill gas 36

Lignin 42, 45, 46, 62, 128, 132, 166, 179, 180, 186

content 128

degradation 62

fraction 179

Modern biomass 30, 50, 65, 160, 161

Photosynthesis 30, 32, 49, 50, 75, 81, 86, 88, 89, 98, 139, 161

Biomass thermochemical conversion processes 162

Biochemical conversion 162

Combustion 21, 23, 39, 40, 50, 58, 62, 64, 69, 86, 111–113, 115, 141–143, 160, 161

direct 21, 33, 34, 40, 50, 97, 161

indirect 30

Electricity from biomass 47

Fischer–Tropsch synthesis (FTS) 36, 50, 161

Gasification 21, 24, 34, 36–38, 40, 45, 50–52, 61, 64–69, 97, 120, 122, 123, 125, 128, 161, 162, 179, 180, 187

steam gasification 120

Hydrothermal upgrading 174, 175

Liquefaction 36, 38, 40, 45, 51, 117, 118, 120, 128, 129–131, 162, 187

cracking 266

direct 36, 40, 128

hydrogenolysis 128

hydroliquefaction 38

indirect 128

of algal cells 127, 129

Pyrolysis 22, 33, 34, 37–41, 45, 49–51, 59, 60, 65, 112, 117–125, 128, 129, 144, 159–162, 174, 175, 178–180

biocrude 17, 38, 39, 51, 107, 108, 160, 175

fast pyrolysis 180

flash pyrolysis 41

hydropyrolysis 41

methanopyrolysis 41

slow pyrolysis 180

ultrapyrolysis 41

vacuum pyrolysis 41

Biorenewable(s) 2, 30, 31, 49, 50, 53, 59, 64, 65, 67, 71, 99, 100, 147, 159, 160, 163, 164, 171, 178, 179

Feedstocks 49,59, 64, 65, 67, 73, 159, 179

algae 98, 101, 139, 146, 173

cellulose 178

corn 21, 22, 33, 52, 75, 98, 101–111, 139, 141, 146, 163, 167, 173, 178, 186–193

corncon 167

corn stover 21, 24

Jatropha 98, 190

maize 108

molasses 21, 23

soybean 21, 75, 98, 108, 139, 141, 190

starch 101, 148, 178

sugar 108

sugar beet 21, 23, 109

sugar cane 22, 23, 75, 98, 109, 139, 186, 190

straw 21, 52, 101, 104, 173

wheat 104, 191

wood 75, 101, 104, 139, 173

Bio-oxygenated liquid fuels 72

Biohydrogen 19, 21, 22, 49, 50, 69, 76, 77, 98, 101, 114, 125, 126, 159, 160, 173

Index

Biogas 19, 22, 32, 51, 61–63, 101, 106, 119, 173
Butane 55

C

Charcoal 40, 42, 72, 106, 121, 180
Coal 2–9, 14, 15, 34, 36, 40, 42, 50, 64–69, 71, 75, 86, 106, 120, 161, 165, 176, 183
  Bituminous coal 8
  Brown coal 8
  Hard coal 7
  Lignite 7

D

Diesel engine 5, 24, 25, 29, 50, 54, 57, 71, 102, 103, 110, 111, 114, 115, 142, 160, 165
Diesel fuel 4, 5, 21, 22, 24–26, 33, 49, 50, 52, 53, 56–58, 67, 68, 70, 100–103, 110, 111, 114–116, 140–142, 146, 149, 153, 159, 160, 163–165, 169, 172, 174, 176, 179, 186
  Alternatives to diesel fuel 114–116
  Oxyganeted diesel fuel 111
  dimethyl ether (DME) 33, 71, 174
  Fischer–Tropsch diesel 24, 101, 173
Dimethyl ether (DME) 33, 71, 174

E

Energy from Algae 97
  Biofuels 99
    bioalcohol 104–116
    biodiesel 102–104
    biomethane 119, 120
    bio-oil 117–119
    hydrogen 120–125
  Liquefaction of algal cells 127–132
Energy sources 1, 2, 3, 11, 13, 17, 31, 140
  Bioenergy 1, 34, 49, 50, 69, 77, 159, 161, 167, 184, 185, 189, 190, 192
  Crude oil 2–5, 21, 38, 52, 55, 56, 108, 162, 163, 168–172, 177, 183, 187
  Fissile 2
    thorium 2
    uranium 2
  Fossil 1–3, 5, 8, 10, 12–14, 16, 17, 21, 24, 25, 30, 32–34, 36, 49, 64, 71, 72, 86, 91, 97, 101,102, 114–116, 119, 121, 149, 159, 160, 183, 166, 183, 186, 193
  Green energy 29–34, 36, 38, 40, 42, 44
  Green power (electricity) 31

Nonrenewable 2, 3
  coal 2–9, 14, 15, 34, 36, 40, 42, 50, 64–69, 71, 75, 86, 106, 120, 161, 165, 176, 183
  petroleum 2–5, 11, 20–22, 24, 26, 49–54, 56–58, 60, 65, 68, 71, 99–103, 106–109, 111, 112, 115, 117, 129, 139–142, 147–150, 153, 159–173, 175, 176, 183
  natural gas 2–4, 8, 9, 14, 19, 22, 25, 32, 34, 36, 52, 53, 64, 65, 71, 72, 102, 106, 109, 110, 159, 170, 175, 176
  gas hydrate 1
  fissile 2
Nuclear 1–3, 8, 10, 16
  uranium 2
  thorium 2
Renewable 1, 2, 3 10–15, 17, 22, 31, 32, 76, 116
  biomass 1, 2, 10, 19–24, 29–39, 43–45, 49–55, 59–61, 63–68, 76–82, 85–87, 97–104, 117–119, 125, 126, 128–132, 139–143, 146–148, 160–166, 173–185
  hydropower 10
  geothermal 1, 2, 10, 11, 15–17, 32
  wind 1–3, 11, 12, 14–18, 21, 31, 32, 87, 153
  solar 1–3, 11, 15–20, 30–32, 49, 75–77, 81, 882, 89, 91, 98, 126, 139, 148, 159, 184
Ethane 55
Ethanol (ethyl alcohol) 6, 21, 23, 24, 30, 33, 36, 37, 42, 43, 45, 49, 50, 52–57, 70–72, 98, 104–107, 109–111, 113, 144, 154, 159, 161, 165–167, 173, 175, 177, 178, 186, 190–193
  Costs 107, 109
  Production 55, 56, 104, 105, 107, 109
    from algae 98
    from biomass 109
  Prices 105, 107, 109
  Properties 54

F

Fischer–Tropsch synthesis (FTS) 36, 50, 161
  Diesel 24, 101, 173
Fischer–Tropsch (FT) liquid fuel 36
Fuel cell 19, 20, 23, 50, 62, 70, 120, 160

G

Gas 2
Gas hydrate 2

198 Index

Gasoline 4–7, 21–23, 33, 36, 40, 50, 53–55,
64, 65, 67, 69, 71, 72, 100, 107–109,
111, 113, 141, 142, 152, 159, 160, 165,
168–170, 172, 174, 178–180
 biogasoline 52, 101, 173, 174
 gasohol 54, 55
 octane number 5, 6, 54, 111, 113, 170
 Otto engine 5, 50, 160
Gasoline-alcohol mixtures 53
Gasoline-methanol mixtures 53
Gasoline-vegetable oil mixtures 53
Glycerine (Glycerol) 70, 141, 142, 145, 165,
176
 Glycerol-based fuel oxygenates 70
  glycerol tert-butyl ethers 70
  propylene glycol 70

H

Hydrogen 1, 8, 16, 21–24, 30, 33, 34, 36, 37,
40–43, 53, 60, 61–67, 69, 70, 98, 105,
120–128, 167, 172, 174–177, 192
 Production 16, 21, 61, 64, 113, 118, 120,
126, 172

I

Internal combustion engine 5, 6, 54, 69, 104,
106, 113, 115, 165

M

Macroalgae 76, 119, 132, 144
 Species 119
  Ascophyllum 19
  Cladophora fracta 144, 145
  Cladophora 19
  Chaetomorpha 19
  Gracilaria 19
  Laminaria 19
  Sargassum 19
  Ulva 19
Methane 30, 33, 36, 51, 61–63, 65, 66, 78,
93, 98, 119, 125, 127, 152, 188
Methanol (methyl alcohol) 24, 25, 30, 36,
37, 41, 49, 53, 54, 56, 57, 60, 61, 71, 72,
102, 106, 107, 113, 141, 144–147, 159,
174, 179, 180
Microalga 49, 75–81, 83, 84, 86–91, 97–99,
102, 103, 118–124, 126, 129, 130–133,
139, 141, 143–153, 185, 190
 Harvesting 79
 Production processes 78, 79, 86, 88, 91,
98, 99, 103, 119, 151

Microalga species 76, 120
 Chlorella protothecoides 144, 145
 Chlorella vulgaris 152

N

Natural gas 3, 4, 8, 9, 14, 19, 22, 23, 25, 32,
34, 36, 52, 53, 64, 65, 71, 72, 102, 106,
109, 110, 159, 170, 175, 176
 Butane 6, 65, 71
 Ethane 65
 Methane 65
 Pentane 65
 Propane 65

P

Petroleum 2–5, 11, 20–22, 24, 26, 49–54,
56–58, 60, 65, 68, 71, 99–103, 106–109,
111, 112, 115, 117, 129, 139–142,
147–150, 153, 159–173, 175, 176, 183
 Asphalt 4, 168, 170, 172
 Crude oil 2–5, 21, 38, 55, 56, 108, 162,
163, 168, 169–172, 177, 183, 187
 prices 38
Pentane 38, 71
Propane 36, 42, 65, 172
P-series fuels 71
 Methyltetrahydrofuran (MeTHF) 33, 71

R

Renewable energy resources 1, 2, 3 10–15,
17, 22, 31, 32, 76, 116
 Biomass 1, 2, 10, 19–24, 29–39, 43–45,
49–55, 59–61, 63–68, 76–82, 85–87,
97–104, 117–119, 125, 126, 128–132,
139–143, 146–148, 160–166, 173–185
 Biorenewable(s) 2, 30, 31, 49, 50, 53, 59,
64, 65, 67, 71, 99, 100, 147, 159, 160,
163, 164, 171, 178, 179
 Geothermal 1, 2, 10, 11, 15–17, 32
 Hydropower 10
 Marine 2, 15, 76, 77, 90, 103, 119, 132,
133, 184
 Solar 1–3, 11, 15–20, 30–32, 49, 75–77,
81, 882, 89, 91, 98, 126, 139, 148, 159,
184
  photovoltaic (PV) 2, 11, 15, 16
 Tidal 1
 Wave 1, 15, 17, 40
 Wind 1–3, 11, 12, 14–18, 21, 31, 32, 87,
153
Rudolph Diesel 165

Index 199

V

Vegetable oils   1, 22, 24, 33, 51, 52, 56, 57,
  70, 78, 90, 98, 100–103, 107, 114, 115,
  117, 140, 141, 144, 146, 148, 149, 160,
  164, 165, 173, 185, 186, 193
  Algal oil   52, 78, 90, 101, 103, 139, 140,
    145, 146, 148, 149, 152, 185, 190
  biodiesel   103, 141, 143, 144, 145, 147,
    150–152

macroalgal oil   145
microalgal oil   78, 152
production   141, 151
Transesterification   141, 145

W

Wood   75, 101, 104, 139, 173

CPSIA information can be obtained at www.ICGtesting.com
233786LV00010B/81/P